WHAT YOUR COLLEAGUES ARE SAYING . . .

Burwell and Chapman provide so many resources that help people see that everyone is a math person. Through vignettes, math examples, talking about mistakes, teaching tips, portfolios, and more, this book has it all!

—Howie Hua
Math Instructor, Fresno State
Fresno, CA

I highly recommend *Power Up Your Math Community* for teachers, instructional coaches, principals, and district leaders! I love the seamless focus on both the classroom and school level, grounded in the implementation of effective instructional practices in a supportive community and developing this as a school-wide commitment. One of my favorite aspects is how the authors balance the true heart of teaching while simultaneously advocating for systemic change. You will love all the helpful features!

—Sarah B. Bush
Professor of K–12 STEM Education, University of Central Florida
Orlando, FL

Innovative, empowering, practical, and inspirational, *Power Up Your Math Community* is a groundbreaking must-read for any educator looking to transform their math education program. Readers will learn how to take a holistic, collaborative, and humanistic approach to professional learning—one that creates an inclusive, enthusiastic learning environment that empowers teachers to create classrooms where learning mathematics becomes a joyous, deeply engaging journey for students and teachers alike.

—Chase Orton
Math Change Agent
Author of *The Imperfect and Unfinished Math Teacher*
Independence, OR

All that is missing is the bow! This book offers a package of professional learning sessions that can be used by all educators who seek to grow their understandings and expand their teaching toolkits for mathematics. When teachers engage in learning and reflect on their practice, students' learning ultimately benefits!

—Susie Katt
K-2 Mathematics Coordinator
Lincoln Public Schools
Lincoln, NE

Power Up will Improve your Math PLC exponentially. Math instructional coaches can stop searching for tools to prompt math growth. It's all here. Instruction, games, student tasks, and reflection prompts.

—Carrie Cutler
Math Coach, University of Houston
The Woodlands, TX

Power Up Your Math Community is an essential resource for schools seeking a year-long, practice-based journey to enhance math teaching and learning. This guide fosters a robust math community through thoughtfully structured monthly themes and essential questions, promoting a culture where students and teachers embrace challenges, learn from mistakes, and grow as mathematicians. It's an invaluable tool for achieving sustained classroom improvement and nurturing a love for math.

—Lisa Ham
Executive Director, Learning Forward Texas
Roanoke, TX

Power Up Your Math Community is a resource for mathematics leaders seeking to enhance math education within their districts and campuses. With its practical guidance, research-based strategies, and emphasis on fostering a culture of continuous improvement, this book provides math educators with the tools they need to elevate math education and empower students to excel in mathematics.

—Nora E. Lugo
Elementary Mathematics Coordinator
San Antonio, TX

This book inspires readers to reexamine their math professional learning plan and provides the much needed road map for quality activities that will build strong professional learning communities (PLCs). Through practical activities that can be used during both PLCs and in the classroom, the authors guide leaders through an easy to follow process that can be implemented immediately. Every school leader should read this and utilize it to create a strong plan for building confident and competent math teachers and students.

—Tobey Realley
Supervisor of Curriculum and Instruction,
Woodbury City Public Schools
Woodbury, NJ

Burwell and Chapman have created a "must have" resource for any leader in mathematics wanting to ensure that both the educators and the students in our classrooms are empowered to work and grow as mathematicians!

—Janet D. Nuzzie
District Intervention Specialist, Pasadena Independent School District
Pasadena, TX

Burwell and Chapman have managed to provide mathematics professional learning at your fingertips in this beautifully written book. It's organized and approachable and feels like I have access to my own math coach. If you are looking to refine math practices in your district, school, or classroom - this book is sure to deliver.

—Rachel Cutler
Elementary Curriculum Coordinator, Great Falls Public Schools
Great Falls, MT

I highly recommend *Power Up Your Math Community* because it provides math educators a step-by-step guide on how to build productive and effective learning environments for all students to see themselves as doers of mathematics and engages them in powerful math habits.

—Jennifer Lempp
Author and Consultant
Alexandria, VA

POWER UP YOUR MATH COMMUNITY

Holly Burwell • Sue Chapman

Foreword by John SanGiovanni

Dedication

To Patty Clark and Mary Mitchell . . .
We are grateful for your servant leadership and advocacy for math education done right.
Your wisdom and compassion for math learners of all ages inspire us every day.

POWER UP YOUR MATH COMMUNITY

A 10-Month Practice-Based Professional Learning Guide, Grades K–5

Holly Burwell • Sue Chapman

Foreword by John SanGiovanni

CORWIN Mathematics

FOR INFORMATION:

Corwin
A SAGE Company
2455 Teller Road
Thousand Oaks, California 91320
(800) 233-9936
www.corwin.com

SAGE Publications Ltd.
1 Oliver's Yard
55 City Road
London EC1Y 1SP
United Kingdom

SAGE Publications India Pvt. Ltd.
Unit No 323-333, Third Floor, F-Block
International Trade Tower Nehru Place
New Delhi 110 019
India

SAGE Publications Asia-Pacific Pte. Ltd.
18 Cross Street #10-10/11/12
China Square Central
Singapore 048423

FSC
www.fsc.org
100%
Paper from well-managed forests

Vice President and
  Editorial Director: Monica Eckman
Senior Acquisitions Editor, STEM: Debbie Hardin
Senior Editorial Assistant: Nyle De Leon
Project Editor: Amy Schroller
Copy Editor: Sheree Van Vreede
Typesetter: C&M Digitals (P) Ltd.
Proofreader: Jennifer Grubba
Indexer: Sheila Hill
Cover Designer: Janet Kiesel
Marketing Manager: Margaret O'Connor

Printed in the United States of America

Library of Congress Cataloging-in-Publication Data

Names: Burwell, Holly, author. | Chapman, Sue, author. | SanGiovanni, John, writer of foreword.

Title: Power up your math community : a 10-month practice-based professional learning guide, grades K-5 / Holly Burwell, Sue Chapman ; foreword by John SanGiovanni.

Description: Thousand Oaks, California : Corwin, [2025] | Includes bibliographical references and index.

Identifiers: LCCN 2024019737 | ISBN 9781071936887 (paperback ; acid-free paper) | ISBN 9781071963050 (epub) | ISBN 9781071963067 (epub) | ISBN 9781071963074 (pdf)

Subjects: LCSH: Mathematics—Study and teaching (Elementary) | Mathematics—Study and teaching (Primary) | Student learning communities.

Classification: LCC QA135.6 .B775 2025 | DDC 372.7—dc23/eng20240720

LC record available at https://lccn.loc.gov/2024019737

This book is printed on acid-free paper.

24 25 26 27 28 10 9 8 7 6 5 4 3 2 1

Contents

Please visit the companion website for
online resources and downloadable materials.
https://qrs.ly/vqfn1s2

Foreword

By John SanGiovanni

There is no argument that a vibrant classroom community is an essential ingredient of successful teaching and learning. Community is the engine for interaction, collaboration, and discussion. It creates opportunity for individuals to experience and consider diverse perspectives that will deepen their own understanding. Community can support and validate new efforts and ideas, which stirs confidence. Community can create belonging through a common pursuit of inquiry, discovery, and acquisition of new skills. Community can be an inclusive place nurtured, and occasionally shepherded, by a knowledgeable, skilled teacher.

Yet, there *are* arguments that community is a missing ingredient in mathematics teaching and learning. Look no further than common utterances like "I'm not a math person" as evidence that many students (and adults) suffer from damaged, if not destroyed, mathematics identities. Where do such notions start? What experiences reinforce negative perceptions? Surely, all of the responsibility for these unproductive dispositions cannot be laid solely at the feet of the math classroom community. But it does cause one to ask, how much of it *is* the outcome of a single classroom community or multiple communities over years of schooling?

From damaged identities and potentially damaging communities springs poor achievement. It is no secret that mathematics achievement in the United States is in a somewhat dire state. Some of this is certainly attributable to a worldwide pandemic in which students were isolated, disconnected, even excluded from learning. (When you think about it, pandemic challenges sound a lot like community challenges, don't they?) But make no mistake, other forces are at work. After all, poor mathematics achievement is not a recent event caused by the pandemic alone.

The mathematics community has responded to poor mathematics achievement results focusing on the content being taught, how it is taught, the number of doses of intervention to provide, and so on. But maybe we should instead be looking more closely at the environment in which mathematics is taught. If great learning comes through engagement, discourse and debate, feedback, support, and perseverance within a rich learning community, then surely there is something to be said about the math community for learning and achievement. Yet for many, that math community has often been fragile, purely transactional, even sterile, denying opportunity and access to the type of community that yields understanding and success.

There are significant challenges to establish and maintain productive mathematics classrooms. Antiquated perceptions of what it means to do math and how math "should" be taught conjure notions contrary to vibrant communities. Beliefs that math is procedural, delivered to students, for them to practice and mimic do not go well with ideas of rich, collaborative communities where ideas are co-developed, analyzed, and refined. Other challenges lie with instructional materials and curriculum standards that do not provide for building and maintaining mathematics community throughout the year. Instead, community is often something "done" or "checked off" the first few days of the year and then put aside for the "real" math. Training for community development is problematic in both pre-service and in-service mathematics teacher training.

The greatest challenge of all may possibly be limiting the notion of mathematics community to the four walls of the classroom or the hour or so a day of instruction. The math community is more than the teachers who teach it. It is everyone! That includes teachers, interventionists, paraeducators, co-teachers, students, and parents. Math community is the school community from classrooms to hallways, the teacher workroom to the conference room, the cafeteria to the principal's office. Beliefs, dispositions, and ultimately actions about mathematics instruction, within those walls and within that hour, are influenced by all of these and much more, including the district offices, homes, supermarkets, and ball fields of the school community. Every one of these mingle to create a mathematics community. But what sort of community is created?

Citing the importance of math community along with some of the challenges is not the same as doing something about the challenges. Ideas for building (or repairing) math community are splattered across journal articles, presentations, and social media. This book, however, does something more. Holly Burwell and Sue Chapman go beyond rhetoric and loosely connected good ideas. They provide a cohesive game plan with clear, doable activities to engage all stakeholders in learning and taking action, so that powerful math communities are something that every student can not only experience but also thrive within.

I encourage you to collaboratively engage with colleagues in your community to study, implement, and reflect on the moves to power up your math community. *Take your time to do it well.* Focus on one idea a month as the authors suggest. Learn from the authors and from the others in your math community, so that the work is not "one-and-done" but continued and strengthened year after year. Your work is the power! That power can change dispositions, change achievement, and even change lives.

Preface

The National Assessment of Educational Progress (NAEP), also known as *The Nation's Report Card*, is a congressionally mandated assessment program designed to measure public and private school student achievement in reading, mathematics, and science in Grades 4, 8, and 12. The NAEP is not intended to identify individual students' achievement but instead is used to evaluate the effectiveness of our nation's schools and education system and to monitor improvement efforts.

Consider these headlines from The Nation's Report Card website page related to the 2022 Mathematics Assessment results:

- Fourth-grade mathematics scores declined across all regions of the country and in 43 states/jurisdictions.

- Fourth-grade mathematics scores declined in 23 of 26 participating urban districts.

- One quarter of fourth-graders performed below NAEP Basic in mathematics—a larger percentage compared to 2019.

- Fourth-grade mathematics scores declined across most racial/ethnic groups; scores declined for male and female students.

Granted, the pandemic contributed significantly to these discouraging assessment results. However, the bigger picture reveals that NAEP mathematics scores have shown little improvement in the last 20 years (see Figure P.1). And the score gaps between Black and Latine students and

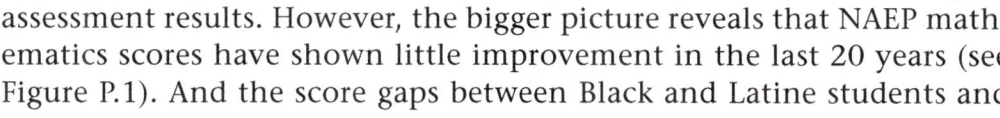

What Does It Mean to Be a Powerful Math Community?

A **powerful math community** is a vibrant group of educators, students, and families, alive with positive energy, efficacy, and a passion for mathematics. Students, teachers, and leaders see themselves and each other as math people and mathematical colleagues. As they engage in rigorous and interesting mathematics tasks, they strengthen their mathematical identities and agency while growing their math understandings and skills. Math is experienced by both children and adults as relevant, empowering, and joyful.

Figure P.1 • NAEP Fourth-Grade Mathematics Average Scale Scores 2002–2022

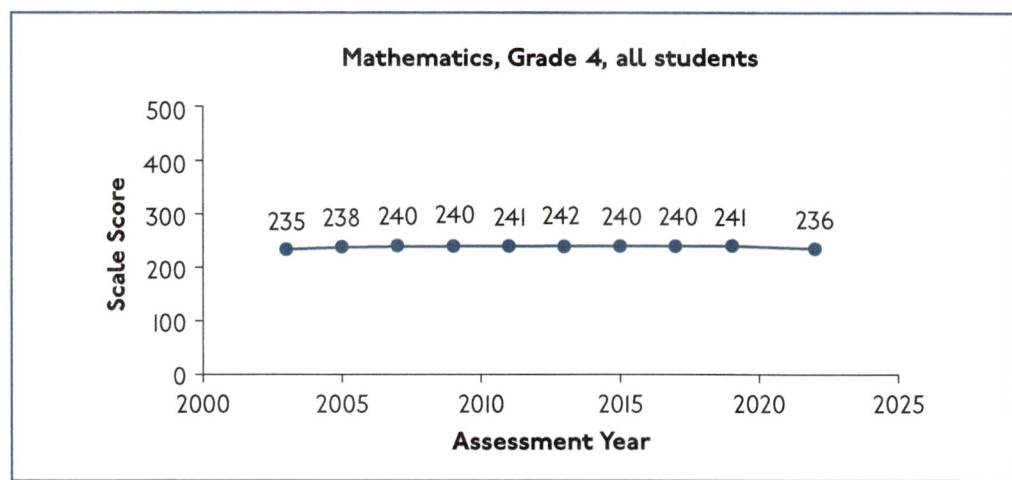

Mathematics, Grade 4, all students

235 238 240 240 241 242 240 240 241 236

their white peers and between female and male students are widening rather than narrowing (The Nation's Report Card, n.d.).

Although the education community has dedicated significant amounts of time, energy, and financial resources to the challenge of improving mathematics achievement, we have very little to show for our efforts.

Why haven't we made more progress? We believe there are three primary reasons:

1. **Expectations for what we teach and how we teach have changed.** The mathematics curriculum and our understanding of best instructional practice have changed in significant ways. Today's curriculum standards charge us with teaching more than just basic arithmetic skills, the only math that many of us learned in elementary school. In addition, we know much more than we used to about how students learn mathematics. We know, for instance, that math learning is supported through social interaction and the use of models and tools (National Council of Teachers of Mathematics [NCTM], 2020). Many of us, however, did not learn math in this way ourselves, and it's really hard to teach in ways that we haven't personally experienced and don't fully understand.

2. **We're fixated on numerical data as a measure of learning.** We have become so habituated to equating test scores with learning that we sometimes lose sight of the true goal of mathematics education: to help students develop mathematical proficiencies that position them to interact skillfully and confidently with the world. Because we forget that numerical data are an imperfect measure of student learning, we haven't fully developed our eyes and ears for observing and interpreting qualitative classroom data (e.g., students' explanations of their mathematical thinking) which often represent student learning more precisely and completely.

3. **We're not yet mindful of mindsets.** We have clear and convincing evidence that students' mathematical identities and agency are the key to their academic success (Aguirre et al., 2013). We know that students' beliefs about their capacity to understand math and the relevance of math to their lives, as well as their habits of mathematical thinking (e.g., perseverance), are critical to learning (NCTM, 2014; SanGiovanni et al., 2020). But we haven't yet committed to intentionally developing these dispositions and "soft skills" as part of our instructional work.

If we wish to finally succeed in improving student achievement in mathematics, we need to think deeply about what mathematics is and how it is best learned. We need to take a careful look at our own relationships with mathematics, our own habits of mathematical thinking, and our personal beliefs about who we are as mathematical beings. *Power Up Your Math Community* will help you to do all of this.

Improving mathematics teaching and learning in the classroom requires that teachers and leaders rebuild their own relationships with mathematics through ongoing practice-based professional learning and collaboration. This important work is most effectively driven at the school level. *Power Up Your Math Community* supports schools in designing, implementing, and evaluating a comprehensive yearlong mathematics instructional improvement plan that allows students, teachers, and leaders to experience mathematics as a joyful activity, to see themselves as mathematically capable, and to effectively grow mathematical proficiency in all math learners.

POWER UP YOUR MATH COMMUNITY WITH PRACTICE-BASED PROFESSIONAL LEARNING

This book is designed to help you and your school engage in **practice-based professional learning** to maximize your students' math learning and strengthen your mathematics teaching and learning community.

Practice-based professional learning is educator learning with the twin goals of strengthening teacher instructional practice and maximizing student learning. It helps teachers sharpen their ability to look at the link between specific teaching actions and their students' learning.

Practice-based professional learning occurs inside or close to the classroom. It is most often collaborative because social interaction supports thinking and energizes the learning community. Therefore, the impact of practice-based professional learning stretches beyond a single classroom. It naturally supports instructional improvement across a team or an entire school.

Practice-based professional learning is teacher driven and inquiry based, designed to address challenges educators uncover in their efforts to support student learning. The process of practice-based professional learning occurs in learning cycles that iterate into new learning cycles (see Figure P.2). Practice-based professional learning is a continuous improvement process that melds with the definition of what it means to be an educator.

Each chapter in this book will guide you and your school through a mini professional learning cycle focused on a specific math habit. The book as a whole is a guide to a larger professional learning cycle designed to power up your mathematics instructional program and your math learning community.

"Most folks don't like math. Much of the anecdotal data we read on social media and hear in real life tells a story of dwindling interest in mathematics combined with an almost contagious distaste for it. It's a subject that often stirs emotions of disdain or resentment – or worse, memories of trauma from their experience in math class."

—Orton (2022, p. 14)

Figure P.2 • Monthly Professional Learning Cycle

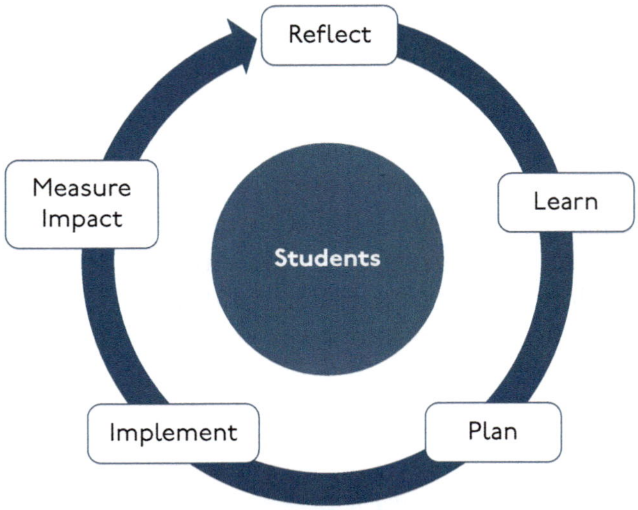

WHO IS THIS BOOK FOR?

By design, *Power Up Your Math Community* has both a classroom-level and a building-level focus. These two focus areas are addressed in tandem because we believe:

1. Teachers are more successful implementing new instructional practices when they have the support of a learning community (MacDonald, 2023; Short & Hirsh, 2023).

2. Student learning is strengthened when a school's mathematics program is cohesive, when there is a school-wide commitment to the use of high-yield instructional practices and the growth of teachers' math content and pedagogical knowledge across grade levels (Karp et al., 2021).

Power Up Your Math Community will help elementary school educators:

- Strengthen their school's mathematics program to provide all students with opportunities to successfully engage in rigorous mathematics learning every day.

- Provide high-quality practice-based professional learning focused on improving math teaching and learning across a school year.

- Build a mathematical community of students, teachers, school leaders, and parents who see mathematics as a lens for understanding and appreciating the world and a way of thinking that allows them to tackle interesting and important real-world problems.

- Plan and implement engaging mathematics learning activities in the math classroom that support students' growing mathematical proficiency while building their mathematical identities and agency.

- Monitor and celebrate growth related to these important aspects of mathematics teaching and learning.

ALL EDUCATORS PLAY AN ESSENTIAL ROLE IN GROWING A POWERFUL MATH COMMUNITY

Power Up Your Math Community is a guidebook for all educators who play a role in improving a school's mathematics instructional program to strengthen students' math learning.

If you are a teacher, this resource will help you . . .

- Grow your students' competence and confidence as mathematicians

- Build your math content and pedagogical knowledge and skills

- Collaborate with team members to strengthen your grade-level mathematics program

You might use this resource in team meetings, with a teaching colleague, or on your own.

If you are an instructional coach, this resource will help you . . .

- Plan coaching cycles in support of individual teachers and teacher teams

- Provide robust practice-based professional learning across a school year

- Deepen your own understanding of math teaching and learning and adult professional learning related to mathematics

You might use this resource with grade-level teams, professional learning communities (PLCs), math vertical teams, and with individual teachers.

If you are a principal, this resource will help you . . .

- Support goal setting and strategic planning to strengthen your school's math program

- Monitor progress toward achieving the improvement goals you identify

- Build your own understanding of research-based instructional practice and curriculum expectations for mathematics

You might use this resource with your leadership team or your entire faculty.

If you are a district leader, this resource will help you . . .

- Support principals, assistant principals, and instructional coaches
- Plan and implement a yearlong professional learning series
- Gain fresh ideas and perspectives about ways to promote instructional improvement in math

You might use this resource in professional learning contexts, district-level meetings, and for campus support.

A COMMITMENT TO EQUITY

Professional learning thought leader Aguilar (2020) stated that every conversation we have in and about schools is a conversation about equity. We agree. We believe that equity conversations and equity work must be a part of every professional learning and school improvement initiative.

Aguilar defined equity in the following way:

> Educational equity means there is no predictability of success or failure that correlates with any social or cultural factor - a child's educational experiences or outcomes are not predictable because of their race, ethnicity, linguistic background, economic class, religion, gender, sexual orientation, physical and cognitive ability, or any other socio-political identity marker. (2020, p. 6)

Everything we do as math educators is an effort to help all students grow into resourceful mathematicians who confidently and strategically leverage habits of mathematical thinking to support their own and others' math learning. Therefore, the goal of equity must drive our every thought and action. It is the heart of our daily work. In *Power Up Your Math Community,* we strive to help teachers and leaders grow their understandings and expand their toolkits of strategies for promoting equity in math classrooms and schools. This commitment to equity runs throughout the book. We also highlight specific equity strategies in call-out boxes titled "Spotlight on Equity" because we believe educators are more effective when they are mindful of the strategies they use to support student learning. We believe that equity work is best supported through collaborative, practice-based professional learning. And so, we encourage you to use the resources in this book to make equity-focused conversations a part of your community norms.

In this book, we strive to use inclusive language. We use the currently accepted terms white, Latine, and Black to honor people's preferences for what they are called. We use the gender-neutral pronoun "they" rather than "he" and "she" whenever possible.

WHY WE WROTE THIS BOOK

We believe school systems should be places where classrooms, leaders, teachers, and students experience joy in mathematics. We want to stand beside you as you create this reality. For too long, we have ignored the research evidence that all human beings can learn rigorous mathematics with understanding. We've allowed the results of imperfect and limited assessments to shape our expectations for students' learning and dictate what students are or aren't exposed to in math class. We save the fun and beautiful part of mathematics for students we label "high" and condemn our "low" students to year after year of mindless drill-and-kill exercises. We've created a culture warped by its preoccupation with identifying students' deficits, blinded to each of our student's innate and ever-growing brilliance. It's time to look at and celebrate our own "mathness," our inborn mathematical nature, and help our students and our colleagues to do the same. It's time for us all to experience math as joyful and relevant to every aspect of our lives. As an educator, your beliefs and actions impact your students' identities and options. We want you to feel empowered to make these important changes. It's time we see all of our students as math capable.

ONE DISTRICT'S INCREDIBLE SUCCESS

In 2023–2024, I (Holly) took a position as a district math coach with Great Falls Public Schools, a school district of 10,000+ students and 700+ teachers. At first, I felt overwhelmed, wondering how I could impact student learning across this entire system. But I believed that if I could help our leaders, teachers, and students begin to feel joy in learning and doing math, we would make a difference. I didn't realize, however, the impact this mindset could have in just one year.

Before I share the students, teachers, and leaders' successes, I think it's important to examine why the work we did is important. We've all heard stories of math phobia, math trauma, and negative attitudes toward mathematics. You can read about this pervasive problem in Vanessa Vakharia's book *Math Therapy(TM): 5 Steps to Help Your Students Overcome Math Trauma and Build a Better Relationship With Math (2025)* and in Lidia Gonzalez's *Bad at Math? Dismantling Harmful Beliefs That Hinder Equitable Mathematics Education* (2023). How can we expect our students to learn, grow, and thrive in mathematics if their math experiences and their teachers' math experiences reflect trauma and negativity? In choosing to focus on math joy, we are focusing on the people and humanness of doing math. When our school communities feel joy and success in learning mathematics, they are motivated to build on that success.

Between 2020 and 2023, Great Falls' growth and achievement assessment data showed a steady decline in students' math achievement at all

15 elementary schools. It was time to ask and answer the question, *"How do we help our students make growth in mathematics?"* Often, in a frenzy to get students "caught up," districts turn to new curriculum resources or computer applications that claim to do it all. Fortunately, based on the advice of our teachers, the district refrained from these actions and chose to put its energy into growing the habits of mathematical thinking in all students. I advocated for also including teachers and leaders in this learning work, knowing that a mindset shift in our educators would benefit our students as well. The beliefs and culture of the school community drives what and how students learn.

The dedicated educators in Great Falls Public Schools rose to the occasion, adopting a vision of joyful mathematics teaching and learning and working together to build positive and powerful math communities within each school. By January 2024, students at all 15 elementary schools were on track for students to achieve one to two years of growth in mathematics. By the end of the school year, ***every single elementary school in our district in grades K-6 met and/or exceeded its instructional improvement goals in mathematics***. Eight out of 15 schools boasted averages in all grades K-6 above proficient, something this district had not seen before. When a community of educators makes a commitment to building positive and powerful math habits and mindsets, students grow and thrive.

It is our wish to you, reader, that this book supports you in building your positive and powerful math community. When you focus on your own relationship with math, the beauty and joy that math brings, and the habits of being a math learner, you and your students will grow and flourish!

Acknowledgments

HOLLY'S ACKNOWLEDGMENTS

When I started consulting years ago, I couldn't think of a better name than "Inspired Mathematics." It was an homage to the many people who inspired me in so many different ways. To my first mentors, Laurie Matteson and Kathy McLean, who if it weren't for them, I would not have been introduced to the brilliant mind of Marilyn Burns. To my coaching mentors who taught me to listen, give grace, and lead with courage: Shelly Kelly, Marni Napierala, Rachel Cutler, Ruth Uecker, and Lindsey Johnson. To my teaching partner who laughs, cries, debates, and aims to seek truth in teaching mathematics: Courtney Francetich. To my Math Solutions mentors who have become some of my best friends and colleagues: Patty Clark, Sandra Coulson, Diane Reynolds, Lucinda Surber, Hannah Pacifici, and so many more. To my friends who listen to me speak about math so passionately and who enjoy so lovingly calling me a "math nerd": Jennifer Duda and Blair Johnson. To my family who fiercely encourages me and patiently supports me as I continue to reach for the stars: my husband Kyle and my mom Kim.

So many incredible math leaders have influenced my career in education. They are dedicated to inclusivity and joy in learning mathematics. Marilyn Burns's work will forever be the reason why I fell in love with teaching math, and it was an honor to be a part of her influence at Math Solutions. I've spent hours on Robert Kaplinsky's website solving problems and sharing lessons with my colleagues. Being a part of the Twitter community has allowed me to learn and grow as a professional, and I highly recommend Howie Hua, Dan Meyer, Fawn Nguyen, Pam Harris, and Jenna Laib. Jo Boaler, Graham Fletcher, and John SanGiovanni's works have contributed significantly to my work as an educator, leader, and learner. Jennifer Lempp offered to be a thought partner and spoke kind words of encouragement and support when I shared I would be writing a book.

Sue has been my rock and leader as I learned how to write a book. Her knowledge, ideas, organization, and patience led us both to write this incredible work. I am honored to have partnered with her.

SUE'S ACKNOWLEDGMENTS

Teaching is a learning profession, and I am deeply grateful to the multitude of people who helped me grow as a teacher and teacher educator. I have learned and continue to learn important lessons from my colleagues, leaders, and students every day. Like stars in the universe, my mentors are far too numerous to name here. We are fortunate to be part of a profession in which fellow learners consistently give so generously of themselves.

I am grateful, too, for the amazing education thought leaders from whom I have had a chance to learn. As an early career teacher, I learned to teach math for understanding and how to listen to students' math thinking from Marilyn Burns. From Jo Boaler, I learned about the importance of paying attention to learners' mathematical identities and relationships with math. Deborah Ball opened my eyes to how teaching mathematics can and must promote social justice. I am proud that these leaders' important work, as well as the work of many other math education leaders, is reflected in our book.

Holly has been an incredible writing and thinking partner. I am so thankful for her expertise and her heart for teachers and students. Her dedication and hard work made this book possible.

****

We thank John SanGiovanni for agreeing to write our foreword. We stand in awe of his work and are honored to have him introduce our book. We thank the Corwin team for helping us transform our vision of a resource to help schools strengthen their mathematics programs into reality. Nyle De Leon, Scott Van Atta, Amy Schroller, and Margaret O'Connor provided expert guidance at every step. Debbie Hardin, our editor, has been our cheerleader since day one. She led this project with passion, curiosity, and kindness and celebrated with us every step of the way. We are grateful to have her be a part of this book.

ACKNOWLEDGMENTS

Tobey Realley
Supervisor of Curriculum and Instruction, Woodsbury City Public Schools
Woodbury, NJ

Jane D. Nuzzie
District Intervention Specialist, K–12 Mathematics, Pasadena Independent School District
Pasadena, TX

Jennifer Lempp
Author and Consultant
Alexandria, VA

Vada Gray
Math Blogger and Advocate, Black Girl Math
Flossmoor, IL

Courtney Francetich
Math Specialist
Great Falls, MT

Carrie Cutler
Math Coach
Clinical Associate Professor, University of Houston
Houston, TX

About the Authors

Source: Lindsey Johnson Photography

Holly Burwell is a teacher, math specialist, learning facilitator, and instructional coach with students and educators in grades Pre-K through 12. She consults through her business Inspired Mathematics supporting math educators across the country and regularly presents professional learning sessions at conferences, including NCTM, NCSM, and the Model Schools Conferences. Holly worked as a math specialist in Great Falls Public Schools to lead and grow math achievement for 10,000+ students. She blogs about math experiences in the classroom on her website at inspiredmathematicsmt.com. Holly is committed to bringing joy to teaching and learning mathematics in schools. You can connect with Holly at hollyburwell@inspiredmathematicsmt.com and @holly_burwell.

Sue Chapman has served students and other math learners for more than 40 years in the roles of teacher, mathematics coach, campus administrator, district curriculum coordinator, and teacher educator. She currently teaches mathematics methods courses to preservice teachers at the University of Houston Clear Lake and provides virtual professional learning and coaching support to teachers and leaders across the country. Sue is the co-author of the book *MathVentures: 33 Teacher-Coach Investigations to Grow Students as Mathematicians* (Math Solutions, 2021) and has written numerous blog posts and articles about coaching, teacher leadership, instructional improvement, and math education. Sue is passionate about building capacity and collective efficacy in educators, teams, schools, and districts. You can connect with Sue at SueChapmanLearning@gmail.com and @SueChapmanLearn.

CHAPTER ONE

Getting Ready to Power Up Your Math Community

Welcome to a yearlong math learning adventure! Each chapter in this book is designed to help you and your students grow a specific mathematical power. Across the school year, we'll be by your side as you power up your mathematical understandings, skills, and habits. As you participate in these mathematics learning activities with students and colleagues, you'll experience mathematics as fascinating, fun, and relevant to real life. You'll develop new eyes for your own mathematical abilities and the power of your math learning community. We look forward to doing and learning math with you!

—Holly and Sue

HOW TO USE EACH CHAPTER

Power Up Your Math Community is a yearlong **practice-based professional learning** guide to strengthen math teaching and learning throughout your school.

In each chapter, you'll find a month's worth of activities and resources to grow your school's mathematics program. These activities and resources are designed to be used in three contexts:

On Your Own	Together With Your Teaching Community	In Your Classroom

Short readings and reflection exercises to introduce the month's learning focus	Ready-to-use professional learning activities for collaborative settings	Actionable steps teachers can take to help students grow as mathematicians
Features:	Features:	Features:
• Let's Do Some Math!	• Since We Met Last • Let's Do Some Math Together! • Building Our Expertise • Let's Try It • Math Talks • Manipulatives and Models Matter • Game Time	• Anchor Lesson • Let's Do Some Math Across Our School! • Teaching Move • Classroom Routine • Station Activity • Literature Connections • Spotlight on Brain Science • Spotlight on Equity • Mathematical Me: Student Journal and Portfolio • Family Newsletter

Images source: istock.com/Fourleaflover

Each month's professional learning activities are designed to be used in sequence but can also be used flexibly in support of the specific needs and interests of your school community.

This book is meant to sit alongside your math curriculum, not as something more to teach but to offer ideas and tools to support you in helping your students to strengthen their math habits, identities, and agency so they can develop the attitudes, dispositions, and mindsets that are prerequisite to high levels of math achievement. Here are suggestions for using the activities you'll find in each chapter.

Suggestions for Using the Chapter Title Page

Chapter Subtitles

On the first page of each chapter, the chapter subtitle is offered in the form of a mantra that can be used to spotlight that month's learning focus. You may choose to post this empowering statement in classrooms, use it to create bulletin board displays, and/or include it in morning announcements or other start-of-the-day routines.

A Note About Monthly Mantras

In the Apple TV series *Ted Lasso*, the coach and namesake of the show inspires his soccer team with witty sayings, compassion, and kindness. Fans of the show will recognize his signature sign posted in the locker room. The single word "believe" serves as a mantra for the players to capture the essence of the mindset needed for team success.

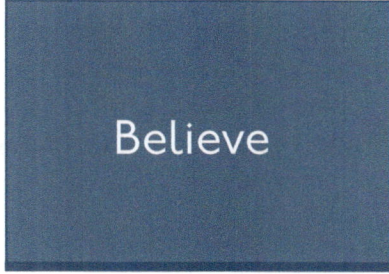

Imagine mathematical mantras in your school or classroom that inspire leaders, teachers, and students to see themselves as learners and doers of mathematics. We invite you to consider each month's chapter subtitle as a mantra, a concise statement of belief that will help grow mathematical mindsets within your school community.

In Season 3 of *Ted Lasso*, the soccer team is shocked to find their beloved Believe sign ripped from the wall. Ted says to his players, "Belief doesn't just happen because you hang something up on a wall. Alright? It comes from in here (heart). You know? And up here (brain). Down here (gut)." The mantra becomes ingrained in the players to guide them ahead with purpose (Lawrence et al., 2020).

This is what we want for our students. We want all students to see themselves as mathematicians in their hearts, their minds, and their guts. When students believe they are mathematicians, they seek out the habits of math learners and, in turn, find that mathematics is fascinating, enjoyable, and rewarding.

Essential Questions

Essential Questions help you and your students to begin thinking about the month's learning focus. They serve as guideposts for learning across the month.

This Month's Focus

This Month's Focus is a concise explanation of the month's learning focus and why it is important for both students and educators.

Mathematical Me: Educator Journal

Regardless of your role in improving mathematics teaching and learning, we suggest that you keep a journal as you read and interact with this book. Doing so will deepen your understanding of the ideas shared. It will also provide a record of your thinking across the year, a learning log, that you can refer back to as you continue your learning journey. You will find educator journaling activities on the first and last pages of each chapter.

Suggestions for Using *On Your Own*

A powerful math community positions all educators (teachers and leaders) as math learners. In this section of each chapter, you'll find short readings and reflective exercises to introduce the month's professional learning focus. You may find it helpful to record your responses to the reflective exercises in your Mathematical Me journal.

Read and think about this section before meeting with colleagues to engage in the professional learning activities in the Together With Your Teaching Community section of each chapter. Alternatively, your school or team might decide to set aside time during a team meeting or professional learning session for team members to do this reading.

Let's Do Some Math!

In Let's Do Some Math! you are invited to tackle an engaging mathematics problem that you will later discuss with colleagues and then use with students. This mathematics problem is also available as a downloadable resource from the companion website (**https://qrs.ly/vqfn1s2**).

online resources

Suggestions for Using *Together With Your Teaching Community*

In this second section of each chapter, you will find ready-to-use professional learning activities for collaborative settings: faculty meetings, professional learning community (PLC) meetings, grade-level meetings, vertical-team meetings, and other professional learning contexts. The first four activities are designed as an hour-long professional learning session for a teacher team or faculty.

Since We Met Last (10 minutes)

Since We Met Last will guide you and your colleagues in sharing and analyzing specific classroom data related to the prior month's professional learning focus.

Let's Do Some Math Together! (10 minutes)

Let's Do Some Math Together! will guide you and your colleagues in sharing and discussing your mathematical thinking from the mathematics task you completed in On Your Own. Reflect with colleagues on your math learning.

Building Our Expertise (30 minutes)

Building Our Expertise describes a collaborative professional learning activity designed to help you and your colleagues transfer what you're learning to classroom practice.

Let's Try It (10 minutes)

Let's Try It will guide you and your colleagues in committing to a simple action step related to the month's learning focus and deciding on the data you will bring to next month's meeting to look at the impact of your actions on student learning.

Doing math together with our teaching colleagues provides in-depth insight into the power and wonder of mathematics. We are a more powerful math community when we grow and learn together!

Suggestions for Using *Additional Professional Learning Activities*

Following the professional learning session outline, you will find three additional professional learning activities to extend teacher learning related to the month's learning focus. These activities mirror recommended classroom teaching practice, building teachers' mathematical understandings while providing a model of recommended strategies to use with students. School leaders and instructional coaches will find these additional activities helpful in preparing monthly professional learning sessions for teachers. Teachers will find this section useful as a guide for personalized professional learning.

Math Talks

Each math talk is an engaging mental math activity or problem designed to promote teacher discussion of mathematical ideas and various solution strategies.

A math talk is a classroom discussion of a mathematical problem or idea. It is brief, five to 15 minutes, and typically takes place at the start of math time. In a math talk, students' mathematical reasoning and explanations take center stage. The teacher uses questioning and other facilitation moves to clarify and deepen thinking, orient students to each other's ideas, and support the group in co-constructing mathematical understanding.

A math talk is a mathematical playground of sorts, a place where students can try out math thinking and build computational fluency in a low-stakes, recess-like environment. As they engage in a math talk, students gain both skill and confidence in mathematical reasoning and in talking about their math thinking while practicing the habits of mathematical learners. They also experience doing mathematics as a joyful endeavor, thus, strengthening their relationships with mathematics and their identities as mathematicians.

Math talks are a powerful vehicle for teacher professional learning. Math talks can be used to instantly engage educators in a fun, learning-focused experience at the start of a team meeting, PLC, or faculty meeting. Math talks with teaching colleagues spark smiles and laughter, so important to educators' emotional well-being and sense of community. Just as this quick learning routine does with students, math talks build teachers' mathematical understandings and reasoning abilities while strengthening their awareness of themselves as mathematical beings.

A Mathematical Question to Consider: If your school spent just 10 minutes at the start of each weekly faculty meeting in a collegial math talk, how might this yearlong investment of 360 minutes or six hours impact educators' math content knowledge, pedagogical expertise, and collective efficacy?

How might this enhanced educator expertise and confidence impact students' math learning and their relationship with math?

Manipulatives and Models Matter

This professional learning experience engages teachers in using a specific math manipulative or model to build familiarity with different mathematical tools and math content knowledge across grade levels.

Manipulatives and other types of mathematical models matter because they are essential supports to math learning (Hiebert et al., 1997; Karp et al., 2021; Tapper, 2022; Van de Walle et al., 2019). According to *Visible Learning for Mathematics* researchers Hattie et al., "Teachers and students can and should use [manipulatives] to make concepts concrete and visible, look for patterns, make connections, and form generalizations. They can likewise be used when constructing viable arguments and critiquing others' reasoning" (2017, p. 170). Manipulatives and models allow students multiple entry points to complex math problems and therefore increase access to rigorous learning. Strategic use of manipulatives and models as learning scaffolds supports equity in the math classroom (Chapman & Mitchell, 2021).

Manipulatives and other mathematical models and tools are spotlighted in the Together With Your Teaching Community section of each chapter as a way of building teachers' representational fluency and their understanding of how mathematical tools support thinking about, talking about, and solving mathematical problems. Many of us had little experience using concrete manipulatives and mathematical tools as mathematics learners in elementary school. When teachers are given opportunities to use mathematical models in support of their math professional learning, their math content knowledge deepens. They become better equipped to use these learning tools effectively in their math instruction and to support all students, K–5, in learning how to strategically choose and use a variety of mathematical models and tools.

Game Time

In Game Time, teachers play a math game to deepen their understanding of mathematical ideas and the month's learning focus. Teachers can then choose to use this same game in their classrooms. Each game is also available as a downloadable resource from the companion website.

Math games are offered as a professional learning experience to deepen understanding of the month's learning focus and to grow educators' pedagogical and math content knowledge. Whenever possible, the games that are shared are appropriate for use across grade levels so they can be enjoyed for learning and used recreationally.

We advocate for the use of math games in all mathematics classrooms because they benefit students in so many ways:

- Math games provide meaningful practice of essential math skills.

- Math games offer students opportunities to engage in mathematical reasoning and communication.

- Math games support the development of social skills.

- Math games strengthen students' mathematical identities and agency.

When introducing a new math game to adults or children, play the game together as a whole group. Once the rules of the game are understood, play can continue with partners. As teachers or students become comfortable with the game, encourage discussion of strategies as a way of surfacing the mathematics in the game. In the classroom, once a game is familiar to students, it can be used as a station or menu activity (Burns, 2023).

The Marilyn Burns podcast *Why play games in math class?* offers additional suggestions for using games to support students' math learning (https:// bit.ly/4c3RrgK).

Suggestions for Using *In Your Classroom*

In the third section of each chapter, you will find actionable steps to help students grow as mathematicians and to strengthen classroom mathematics learning communities. You can use this section to choose learning activities and teaching moves to implement on your own or together with colleagues. This section is a resource of specific teaching strategies to strengthen mathematics instruction in individual classrooms and across the school.

Anchor Lesson

An Anchor Lesson develops student understanding of the month's learning focus.

Let's Do Some Math Across Our School!

This section includes a mathematics problem related to the problem you explored on your own and with colleagues. If possible, display students' problem-solving work outside your classroom or in a public area so that other students, teachers, and school visitors can admire the mathematical brilliance that exists across the school community. You might organize "field trips" for students to look at work from other classes or grade levels, perhaps leaving sticky notes with positive feedback, as a way of helping students to see each other as fellow mathematicians and mathematics as an activity that people of all ages enjoy and want to talk about.

Teaching Move

Beginning in Chapter 3—September, you will find suggestions for specific teaching moves to support the month's learning focus.

Classroom Routine

Beginning in Chapter 3, you will find suggestions for classroom routines to support the month's learning focus.

Station Activity

Beginning in Chapter 3, you will find a suggested station activity that students can do independently to deepen understanding of the month's learning focus. Each station activity is also available as a downloadable resource from the companion website.

Literature Connections

Literature connections offers suggestions for children's literature related to the month's learning focus.

Spotlight on Brain Science

Spotlight on Brain Science offers a simple explanation for students of an idea related to the month's learning focus and brain research. The student reading for each Spotlight on Brain Science feature is available as a downloadable resource from the companion website.

Spotlight on Equity

Spotlight on Equity gives specific suggestions for teaching moves related to the month's learning focus to build equitable math classrooms.

Mathematical Me: Student Journal and Portfolio

The student Mathematical Me activities offer journal prompts and suggestions for student portfolio work to document the learning that takes place across the year. A simple math journal template is available on the companion website for duplication, or you may decide to have students use composition books or spiral notebooks for their math journals.

File folders or 18″ × 24″ sheets of construction paper folded in half can serve as math portfolios. We recommend that students' math journals and portfolios be stored in a central location in the classroom rather than in students' desks to preserve these important learning records across the year.

Younger students who are not yet writing may draw a picture in response to journal prompts. Kindergarten and first-grade teachers may also want to partner with a buddy class of older students who could visit once a month to scribe younger students' journal and portfolio reflections.

Family Newsletter

The Family Newsletter is a one-page explanation of the month's learning focus for families that also provides ideas to use at home. Each Family Newsletter is also available as a downloadable resource from the companion website.

Suggestions for Using *Checking In On Our Learning*

This final section of each chapter takes readers back to the month's Essential Questions and the opportunity to reflect on the learning that has taken place. It also previews next month's learning focus.

Downloadable Resources

To support your work, a variety of downloadable resources are available on the *Power Up Your Math Community* companion website at **https://qrs.ly/vqfn1s2**.

PREPARING FOR YOUR LEARNING ADVENTURE

The educator and student learning activities in this book are designed to make it easy for you and your team to get started on this exciting learning adventure. Here are a few logistical details you'll want to think about before beginning.

How Will You Make Time for Monthly Professional Learning?

Time is a limited resource in schools. The monthly professional learning sessions are designed to take just one hour with this reality in mind.

(Note that additional professional learning activities provide options for extending this.) You'll want to make decisions upfront about whether the monthly professional learning will take place with your entire faculty, in grade-level teams, in a vertical math team, or in another setting.

The professional learning activities in this book will deepen educators' understanding of mathematics and mathematics education. But, just as important, these regular professional learning sessions offer your teams and faculty the chance to share what they're trying in the classroom and how their actions are impacting student learning. These discussions position your teaching community to co-construct important professional knowledge that can maximize mathematics learning across your school.

How Will You Introduce the Power Up Your Math Community Learning Initiative to Your School Community—to Staff Members, Students, and Parents?

This year will be an important event in the life of your school. Community members will have the chance to come together to create a shared definition of what mathematics is and to strengthen their personal relationships with mathematics. You'll want to let everyone know what you'll be doing and why it's important.

In an early faculty meeting, you might read and discuss the definition of a Powerful Math Community and the rationale for this important work offered in the Preface. You'll want to tell students on the first day of school that every member of the school community—students, teachers, and families—are all mathematicians and that you'll all be growing your mathematical powers across the year. You'll want to tell parents about the importance of this community-wide learning work. Let them know that you'll keep them posted with monthly newsletters and that they can look forward to an end-of-year celebration of their child's math learning.

How Will You Keep the Math Conversation Going?

Growing a team of educators who are passionate about improving mathematics teaching and learning will be essential to the success of your Powerful Math Community. Building your school-wide mathematics program can feel daunting; the collective efforts and shared commitment of a team of teacher-leaders will be key to sustaining this important work. Create routines and structures to keep the math conversation going within your school community to build a school culture for continuous improvement of your mathematics instructional program.

> Regular professional learning sessions offer your teams and faculty the chance to share what they're trying in the classroom and how their actions are impacting student learning. These discussions position your teaching community to co-construct professional knowledge that can maximize mathematics learning across your school.

FAST FORWARD TO THE END OF THE SCHOOL YEAR

Planning Ahead for Your End-of-Year Math Celebration

"Recognizing and celebrating success contributes to building a culture for learning."

—Killion et al. (2023, p. 169)

As your school year comes to an end, it will be important to pause and celebrate the learning that occurred. In the Let's Do Some Math Across Our School! section of Chapter 11—May, you'll find suggestions for a school-wide end-of-year celebration of math learning together with families. This celebration could take place during a family math night and/or student-led conferences. Students' Mathematical Me journals and portfolios will be important learning artifacts during this celebration as will the classroom anchor charts that teachers and students create throughout the year. It's a good idea to make tentative plans now for this celebration. Put the date on your school calendar and decide how you will collect and curate evidence of student and teacher learning each month.

Planning Ahead for Next Summer

If possible, make plans now for when and how you will bring your faculty or leadership team together at the end of the school year to reflect on your learning and begin planning for the following school year. You might dedicate time during a professional learning day or staff retreat. Finding a way to involve all teachers in reflecting on and planning for professional learning in support of students' math learning will embed a continuous-improvement mindset within your school culture to continue powering up your math community.

The downloadable resources available on our companion website include a bonus chapter resource called "Chapter 12—This Summer's Focus: Powering Up for a New School Year!" This resource will support you and your teaching community in preparing for this important reflecting and planning conversation. In addition, the Appendix at the end of this book offers a short list of outstanding professional books to support your school's next steps in powering up mathematics teaching and learning.

CHAPTER TWO

August
We Are All Math Learners!

ESSENTIAL QUESTION

- How do I know I am a math learner?

THIS MONTH'S FOCUS

In a powerful math community, leaders, teachers, and students know they are all math learners and they see the value that each person brings to a math classroom. They see the beauty of mathematics all around them and seek opportunities to learn more.

Teachers and leaders elevate all students to be seen as the mathematical thinkers that they are. Students find joy and beauty in doing mathematics.

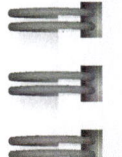

MATHEMATICAL ME: EDUCATOR JOURNAL

Take five minutes to respond to the essential question. You will have a chance to reflect on this same question and to have students respond to this question at the end of the month.

ON YOUR OWN

LET'S DO SOME MATH! MATH IS A PART OF ME

Consider for a moment the impact you have had on countless students. The interactions you have each day with teachers and students potentially change the trajectory of their lives. Determine a way to calculate the number of students you have influenced in your career. How could you quantify any future impact you have on students? (Available as a downloadable resource on the companion website at **https://qrs.ly/vqfnls2.**)

Holly's example (see Figure 2.1):

Camp counselor: 7 years × 40 students per year = 280 students

Teacher and coach: 9 years × 25 students per year = 225 students (see Figure 2.1)

National coach and consultant: Approx. 300 districts: 500 teachers × 25 students for each teacher = 12,500 students

Figure 2.1 • Holly's Summer Math Camp Students

Why This Focus

For too long, it has been believed that people are either born being good at math, having a "math brain," or not. Research has proven time and again that this simply is not true. In her book *Mathematical Mindsets,* Jo Boaler (2015) wrote about the work of neuroscientists who have found that our brains have neuroplasticity—in other words, they strengthen (or weaken) and change throughout our entire lives. This neuroplasticity is also true with learning math. All people have the potential to learn math at high levels. So how can we convey this message to leaders, teachers, and students in our schools?

Teaching and learning math in school involves a balance between math content and math practices. These two components of our mathematics curriculum are intertwined.

Image source: istock.comRockantansky

Although the content is specific to the grade, such as 1-digit by 4-digit multiplication in fourth grade, the math practices cross all grade levels and guide students on the "how" of doing mathematics. The math practices are the behaviors one uses while doing math. All students, teachers, and adults use the math practices daily.

The math practices grew from two influential forces in mathematics education: the National Council of Teachers of Mathematics's (NCTM, 2000) Process Standards, which include problem solving, reasoning and proof, communication, representation, and connections, and the National Research Council & Mathematics Learning Study Committee's report, *Adding It Up* (2001), which lists proficiencies that mathematics educators should develop in their students. These traits include adaptive reasoning, strategic competence, conceptual understanding, procedural fluency, and productive disposition. All state and provincial mathematics standards include math practices. In many states, they are referred to as the **Standards for Mathematical Practice (SMPs)** (National Governors Association Center for Best Practices & Council of Chief State School Officers, 2010). For the remainder of this book, we will reference the Standards for Mathematical Practice and refer to them as SMPs.

Here's the list of Standards for Mathematical Practice (SMPs):

1. Make sense of problems and persevere in solving them

2. Reason abstractly and quantitatively

3. Construct viable arguments and critique the reasoning of others

4. Model with mathematics

5. Use appropriate tools strategically including mental math

6. Attend to precision

7. Look for and make use of structure

8. Look for and express regularity in repeated reasoning

Table 2.1 lists examples of teacher and student behaviors associated with each SMP. In a powerful math community, leaders, teachers, and students actively strive to strengthen their use of the SMPs.

Table 2.1 • Examples of the SMPs in Action

STANDARD FOR MATHEMATICAL PRACTICE	TEACHER EXAMPLES	STUDENT EXAMPLES
Make sense of problems and persevere in solving them	Read problems aloud and ask students to turn and talk about what they heard.	Use strategies to make sense of problems: • Reread • Talk with someone about what you know • Act it out
	Read the problems and ask students to visualize or picture the scenario. Share out.	
	Allow students to work individually for two minutes, followed by one to two minutes of sharing with a partner.	
	Bring the class back together after five to seven minutes to share different strategies.	
	Give students strategies for working through challenging problems: • Reread • Talk with someone about what they know • Act it out	
Reason abstractly and quantitatively	Make time for students to estimate the answer to a problem before solving the problem.	Ask, "What quantities will I be working with in the problem? What is a range of answers that seem reasonable?"
	Ask questions to help students think about the reasonableness of their answers.	Think about a reasonable estimate for a solution to a problem.
		Consider what operations will be used and why.

STANDARD FOR MATHEMATICAL PRACTICE	TEACHER EXAMPLES	STUDENT EXAMPLES
Construct viable arguments and critique the reasoning of others	Provide sentence stems for students to communicate their solutions.	Share solutions and the steps and strategies used to get there.
	Make mathematical and academic vocabulary visible in the classroom for students to use.	Share answers whether correct or incorrect.
	Expect students to share their math thinking.	Justify conjectures.
Model with mathematics	Expect students to show their work in various ways, including but not limited to drawings, charts, tables, equations, graphs, written explanations.	Use multiple representations when sharing work.
	Help students connect strategies and concepts.	Look for connections between strategies.
		Find connections between strands of mathematics.
		Use tables, charts, graphs, equations, and/or written words.
Use appropriate tools strategically including mental math	Make a variety of tools available for students to select and use.	Select tools that support problem solving.
	Make time for students to explore with manipulatives.	Use mental math.
	Make time for students to practice mental math.	Determine when certain tools may or may not be appropriate and helpful for solving problems.
Attend to precision	Expect students to use mathematical vocabulary when communicating or writing about their work.	Use precise language when communicating either orally or in writing.
		Check solutions for precision.
		Use estimates to check for reasonableness and refine solutions.
Look for and make use of structure	Develop numerical patterns and ask students to share what they notice (e.g., write a series of related problems).	Look for patterns.
	Use developed patterns and ask intentional questions for students to make sense on their own.	Look for connections between concepts.
Look for and express regularity in repeated reasoning	Use patterns generated to have students make conjectures about what might always be true.	Test theories and derive conclusions.

online resources 🔍 Available as a downloadable resource on the companion website.

Tapping Into Your Experience

Think back to your math learning experiences as a student. What did math class feel like? What sights and sounds stand out to you?

Create a two-column chart. List similarities on the left and differences on the right. Review the list of math practices above and write down any similarities or differences to the math practices you experienced as a learner.

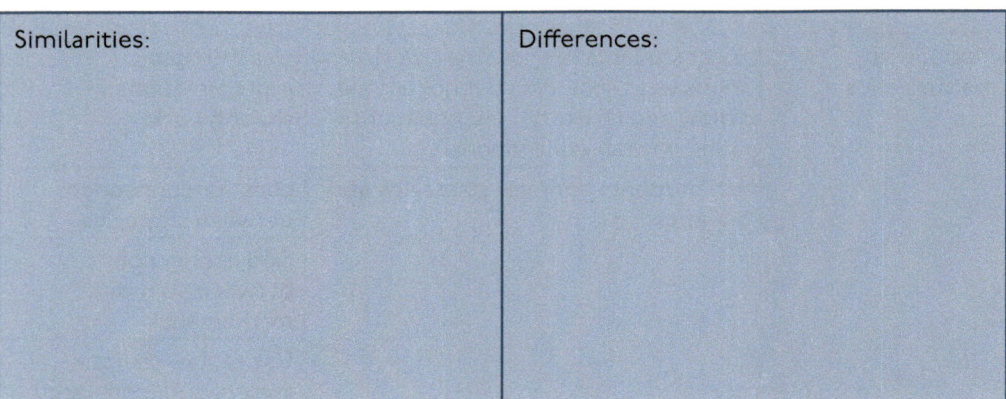

Similarities:	Differences:

After creating and reviewing your list, consider the following questions:

- What experiences do I want to take into my math teaching and leading of math learners?
- Which practices feel unfamiliar and may need more attention? How might I familiarize myself with them?

Habits of Mathematically Powerful People

The SMPs provide teachers with an important focus for their instruction and define essential learning behaviors for students. They can be somewhat difficult for students to understand. In this book, we have chosen to use a student-friendly framework of math practices that we call **habits of mathematically powerful people** (see Figure 2.2). The SMPs and the habits of mathematically powerful people are deeply connected learning practices that grow students as mathematicians.

In the following chapters, we will deconstruct the habits of mathematically powerful people and share practical ways to grow the habits in ourselves and our students.

This chapter offers you and your math teaching community a wealth of resources for introducing students to the habits of mathematically powerful people.

Leaders, teachers, and students need to recognize the mathematical power within themselves and seek to continuously build their own math habits. Throughout this book, we will share personal stories that may be similar to

Figure 2.2 • Habits of Mathematically Powerful People

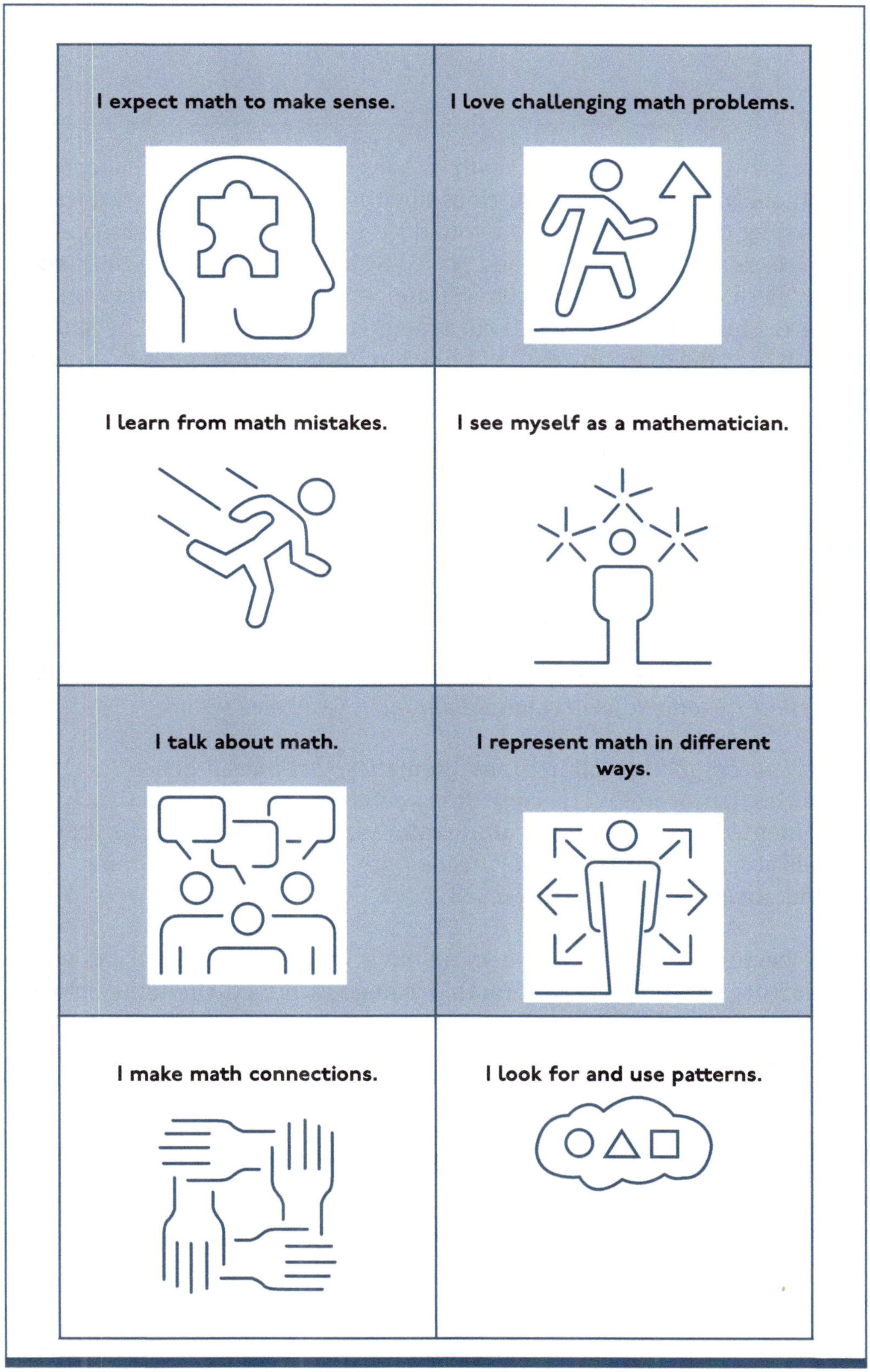

I expect math to make sense.	I love challenging math problems.
I learn from math mistakes.	I see myself as a mathematician.
I talk about math.	I represent math in different ways.
I make math connections.	I look for and use patterns.

Images source: istock.com/Fourleaflover

 Available as a downloadable resource on the companion website.

"We cannot expect students to succeed in achieving current academic standards unless we are at the same time intentionally attending to the competencies that undergird the academic expectations."

—Markowitz and Bouffard (2020, p. 12)

or different from your own mathematical experiences. Our goal is to help you reflect on the experiences that shaped your math identity and consider how we can help our students strengthen their math identities.

Holly's Math Story:

My husband, Kyle, was given a math message in middle school that changed his outlook and belief about himself as a math learner. After not fully engaging or listening during math class, he found his homework to be a bit of a challenge. He sat with it for a while and played with the math until he could come up with a strategy that led to a correct solution. He finished his homework and felt successful that he had figured out a method on his own. When he returned to math class the following day, he turned in his homework feeling confident, especially since he could verify correct answers in the back of the book.

Days later, he received his homework assignment back to find the entire page marked incorrect because he hadn't solved problems using the strategy shown by the teacher during class. At this point, Kyle decided he wasn't really a math person because he couldn't do it "right."

I reflect on this experience, wondering how someone can conclude that math is not a part of them or their lives. Unfortunately, this message has become so pervasive. In her book *Bad at Math? Dismantling Harmful Beliefs That Hinder Equitable Mathematics Education*, Gonzalez wrote:

> The acceptance of failure in mathematics, just like all belief stories, permeates our society. It is perpetuated in the media and cemented in our popular culture. More troubling, however, is that this acceptance finds its way inside our classrooms, boardrooms, and government agencies. (2023, p. 3)

At the beginning of the year, we set up our school spaces to be places where all students see themselves as math learners. Although the table below is not exhaustive, teachers and leaders can use these ideas to think about what their spaces might look and sound like.

Physical Space	Is there space in the room to display our students' mathematics work?
	Are there math tools visible and easily accessible to our students?
	Can our students easily collaborate and speak about math?
Language	Do we speak positively about mathematics?
	Are we referring to all students as math capable?
Attitude	Do we make time for joy in mathematics? For example, do we meet daily at a special place in the room for math talks or to play a math game?
	Are we excited when we talk about math and engage in math?

When we were young, we understood what *half* was after learning that we must share a cookie with our sibling. When we became drivers, we noticed our speed and time to move from one distance to another. As adults, we divide our monthly income by all of our expenses. Although these examples demonstrate practical mathematics, I often wonder how many people see the mathematical beauty when looking at the spirals of a flower or the shape of a flock of geese that fly overhead. Do you notice the patterns of how people sit in meetings or where they sit in a restaurant? Mathematics is woven seamlessly through our lives.

How can our messages help students see themselves as math learners? What if our default was to assume that all people, including ourselves, are capable math learners? A **self-fulfilling prophecy** is the notion that what we believe affects our actions, and those actions confirm our original beliefs (Gonzalez, 2023). Gonzalez spoke to this idea, saying, "How we see or do not see ourselves as doers of mathematics affects our ability to engage in mathematical work and to achieve success in the discipline" (p. 89).

> Mathematics is woven seamlessly through our lives.

In Figure 2.3, we can see how our beliefs and actions are interconnected. What we do, our actions, impact the beliefs other people have about us. As a result, others may take action that reinforces a belief about ourselves.

Figure 2.3 • Model of a Self-Fulfilling Prophecy

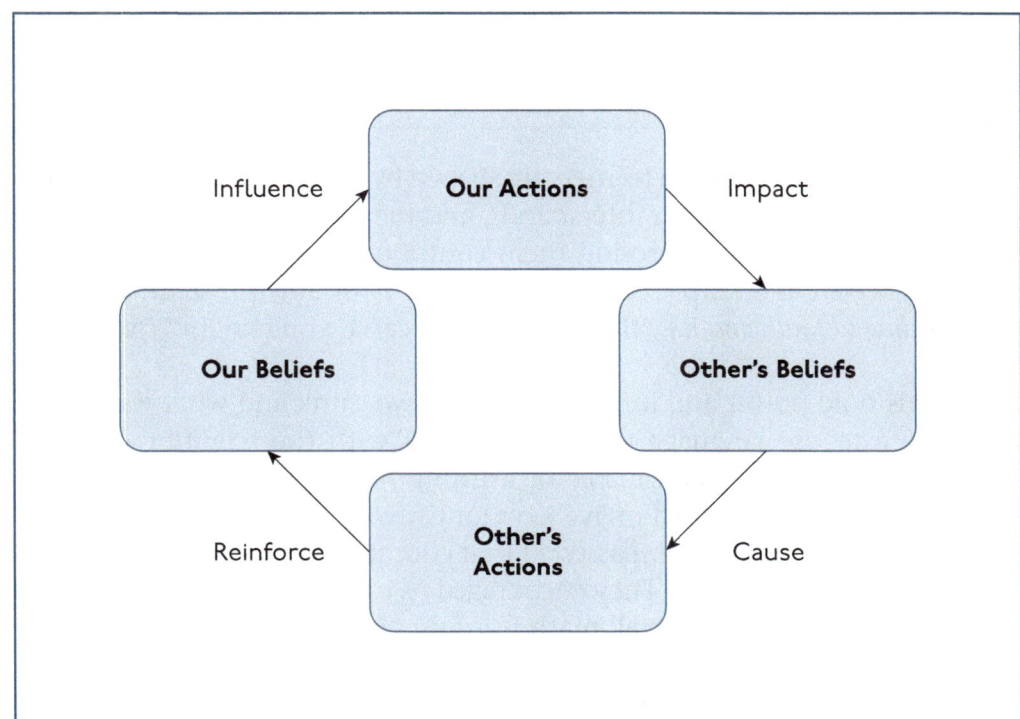

Kyle used to tell me that he wasn't really a fan of mathematics. He didn't see how it was relevant or useful to his life. When we go back to his experience in school, we can see the self-fulfilling prophecy that brought him to this belief. After Kyle turned in homework with his unique strategy for solving his math problems, the teacher marked the answers incorrect and reinforced Kyle's belief that he wasn't a math thinker. Interestingly, he chose a career in technology and works as a database administrator, where he uses math to analyze and extract data. One of his hobbies is woodworking, where his knowledge of mathematics is extensive and necessary for creating projects. It took him years to trust his mathematical creativity and knowledge he now uses daily.

The language we use, the attitudes we hold, and the beliefs we have about ourselves and our students have the potential to positively impact thousands of students over our career. If a student comes to class with a different strategy, we can choose to see that as an opportunity to be curious. We might ask the student to share how they came up with their strategy and then ask them to share it with the class. The student would believe their contribution was interesting and worthwhile, which may encourage them to see themselves as a math learner. In the future, that student will continue to look for unique solutions to problems.

> The language we use, the attitudes we hold, and the beliefs we have about ourselves and our students have the potential to positively impact thousands of students over our career.

In the end, we are all math learners. In building our classroom communities, we have the potential to influence our students' attitudes and beliefs about mathematics. They listen to how we speak about math, watch our body language, and pick up on our enthusiasm for the math we teach. Our students' view of what it means to do math is highly influenced by their experiences in the classroom each day.

Using Strengths-Based Language

Children's identities as math learners are shaped by their experiences in school. The content they learn, their interactions, and the messages they receive from their peers and the adults around them contribute to how they view themselves. As Kobett and Karp (2020) stated in their book *Strengths-Based Teaching and Learning in Mathematics*, "If you think they can't, you're right" (p. 3).

Our words hold power and influence the way we think and what we do. We can choose to use language that is congruent with the scientific evidence that all human beings are capable of learning high-level mathematics with understanding. In their article "Five Keys for Growing Confident Mathematics Learners," Rhodes et al. (2023) asserted that educators should stop referring to students as "high" or "low." They encouraged us to use language that communicates our belief that we are all math learners: "Instead of having 'high' and 'low' students, the classroom contains mathematicians, each with a different and equally powerful method of making sense of ideas" (p. 10).

Here are some examples of educators using **strengths-based language**:

- These students *can* learn math! Let's build from what they already know.

- My students *can* learn math at high levels. They can think and problem-solve.

- They *can* do this part of the problem. What questions will we ask to support them with the part of the problem they aren't sure of?

- How can we incorporate this student's strengths into the lesson?

Consider the language currently used in your school to describe students as math learners. What strengths-based language might you and your colleagues begin to use? How will this language help your students see themselves as mathematicians?

TOGETHER WITH YOUR TEACHING COMMUNITY

Since We Met Last (10 minutes)

This first month of the school year, you and your colleagues are starting your yearlong journey toward becoming a powerful math community. Each of us brings a unique experience base and wealth of knowledge that strengthens our collective team of math leaders. For this month's data review, ask your teachers to bring a math lesson they taught last year that engaged students and had positive learning outcomes.

Use the following protocol to share and reflect on this data with your math teaching community:

- Take two sticky notes and record the following, one on each note: Why was the lesson engaging for students? What made the lesson a positive experience for students? (5 minutes)

- Find a partner and take turns sharing a brief overview of your lesson and your two sticky notes. (5 minutes)

- If time allows, repeat the process with other partners.

Let's Do Some Math Together!

Share your strategy for calculating the number of students you have impacted in your career.

Discuss:

1. How did you approach the problem?
2. What did you need to consider in your calculations?
3. Did your number surprise you? Why?

With your math community, debrief the experience. Use the following questions to guide the conversation.

*   What were some similarities and differences between your strategies and others?
*   Did the number surprise you? Why or why not?
*   How might this number make you think about your future impact on students in mathematics?
*   Why is this number important?

Building Our Expertise (20 minutes)

The math lessons everyone shared that engaged students and had positive learning outcomes likely included many aspects of the SMPs.

Spend three minutes individually reviewing your lesson to identify places where students had the opportunity to engage in the SMPs.

Next, talk in partnerships about what you found.

Share your insights as a whole group and consider this question together:

*   As we think about our continued work to become a powerful math community, how can we orchestrate mathematical experiences that are filled with the math practices?

Review the list of examples of the SMPs in action (Figure 2.1). Choose one or two actions that you can test out in your learning community.

Building Our Expertise (10 minutes)

The SMPs detail how students interact with mathematical content. They are observable behaviors that you can explicitly teach, practice, look for, and assess. The *habits of mathematically powerful people* outlined in this book are beliefs and mindsets that underlie the SMPs.

With your community of learners, lay the SMPs and the habits of mathematically powerful people side by side and think about their similarities and differences.

HABITS OF MATHEMATICALLY POWERFUL PEOPLE	
I expect math to make sense.	**I love challenging math problems.**
I learn from math mistakes.	**I see myself as a mathematician.**
I talk about math.	**I represent math in different ways.**

(Continued)

Images source: istock.com/Fourleaflover

(Continued)

I make math connections. 	**I look for and use patterns.** 

Images source: istock.com/Fourleaflover

 Available as a downloadable resource on the companion website.

Standards for Mathematical Practice (SMPs)

I.	Make sense of problems and persevere while solving them	2.	Reason abstractly and quantitatively
3.	Construct viable arguments and critique the reasoning of others	4.	Model with mathematics
5.	Use appropriate tools strategically	6.	Attend to precision
7.	Look for and make use of structure	8.	Look for and express regularity in repeated reasoning

As a group, share your insights and use the following questions to guide your discussion:

- How are the SMPs and habits of mathematically powerful people similar? Different?

- What are the implications for teaching and learning in using both the practices and habits?

 Available as a downloadable resource on the companion website.

Let's Try It (10 minutes)

With your teaching community, consider which of the SMPs or habits of mathematically powerful people you would like to make your focus. Choose a student example from Figure 2.1 and consider one of the options below to collect data. Record your data collection plan in your planner or calendar, and bring the data to next month's meeting to share and discuss.

OPTION 1: NOTICE AND WONDER

Provide students with a math task to work on in pairs or small groups. Notice what students are doing related to the math practice that you selected. Take anecdotal notes that you can share with your colleagues during next month's meeting.

For example, if you chose the math practice of making sense of problems, what do you notice about how students are making sense of the problem? Are they seeking advice from others?

OPTION 2: ANALYZE STUDENT WORK

Take a sample of one or two pieces of student work to share with your colleagues. How does this student's work reflect the SMP or habit of mathematically powerful people that you are focusing on?

For example, if you chose to focus on the math practice of reasoning abstractly and quantitatively, did students provide reasoning for the solutions they found?

OPTION 3: INTERVIEW STUDENTS

Some practices require the teacher to hear students talking about what they are doing. For example, the SMP of justifying your thinking may be better understood through student conversation.

Select a couple of students, one at a time, and ask a couple of questions related to the practice you chose to study. For example, if you chose the SMP of making sense of problems, you might ask the following:

- How did you make sense of this problem?
- Tell me about your strategy for solving the problem.

ADDITIONAL PROFESSIONAL LEARNING ACTIVITIES

This month's additional professional learning experiences are opportunities to explore the SMPs. In these instances, teachers experience the openness of math and practice making sense, reasoning, justifying, using tools, and looking for patterns.

Math Talks: Dot Talk

This month's math talk is a dot talk fitting perfectly into the beginning of the school year as we are building our math classroom communities. This low-stakes math talk encourages learners to authentically practice the SMPs and the habits of mathematically powerful people.

A dot talk starts by showing an arrangement of dots. Students are asked to share how many dots they see and how they see them. They are encouraged to notice and celebrate the varied ways that others solve this simple problem. To watch a dot talk in action, visit youcubed.org as Jo Boaler takes students through a visual dot card number talk (https://bit.ly/3PxVWqz).

Together with your teaching community, give the dot talk a try. You may use the image below or you can find other dot images from a variety of sources online. *Number Talks* by Parrish (2022) includes a large variety of dot talks to use with students.

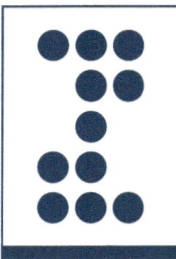

online resources ↖ Available as a downloadable resource on the companion website.

Discuss the experience:

- How did you experience the different SMPs when we did the dot talk?

- How did your use of these practices support your thinking?

- How could you use dot talks in your classroom?

Manipulatives and Models Matter: Connecting Cubes

Mathematics is a beautiful and open subject in which we explore, seek answers to our questions, and make sense of the world around us. When we see math in this way, we play with it, find ways to represent it and communicate through writing, recording, and speaking math. Manipulatives are just one way we can explore our mathness.

One SMP—Use Appropriate Tools Strategically Including Mental Math—and a habit of mathematically powerful people—"I represent math in different ways"—encourages students to use tools to make sense of math. For this month, teachers explore connecting cubes and their versatility in allowing students to think about mathematical concepts.

Try this experience with your math teaching community.

Working in small groups, take a number of connecting cubes that represent something about your group. For example, you might take cubes to represent the total number of pets owned by the group or the total number of years of teaching experience for all group members.

- Share with the whole group the number of cubes your group took. Ask other groups to guess what the number represents.

- Take turns sharing and connecting the quantity to context.

Consider how you might adapt this task for the students in your classrooms.

- Students might contribute to a display titled "Our Classroom in Numbers"

- Students might write about the number or numbers generated starting with the sentence stem "The number represents our classroom community because . . ."

Game Time

This month's game involves mathematical reasoning. It is easily accessible to all students, and the rules are simple. However, finding a strategy to win is more challenging. The more students play, the more they will begin to articulate strategies to outwit their opponents.

ODD NUMBER WINS

Materials:

2 players

Connecting cubes or color tiles

Challenge: Be the player with the odd number of cubes at the end of the game.

Game directions:

1. Place 15 cubes between both players.

2. On your turn, take one, two, or three cubes from the shared pile.

3. Set your cubes in front of you.

4. Play moves back and forth until all the cubes have been taken.

5. The winner is the player who finishes with an odd number of cubes.

Discuss the experience:

- How does this game help all students see themselves as math thinkers?

- What math habits would students engage with in playing the game *Odd Number Wins*?

 Available as a downloadable resource on the companion website.

IN YOUR CLASSROOM

Holly's Story

Early in a school year, I came into a math classroom to partner with a teacher who wanted to see her students engage with an unfamiliar math manipulative. Students were going to use Cuisenaire rods (shown below) as a jumping-off point to learn about the math norms of using manipulatives in the class, and the teacher would use this opportunity to formatively assess students on their foundations of fraction knowledge. The fourth-grade teacher, Mrs. Yadev,[1] asked if I would lead the classroom in their learning experience so she could make anecdotal notes on what students were saying and doing.

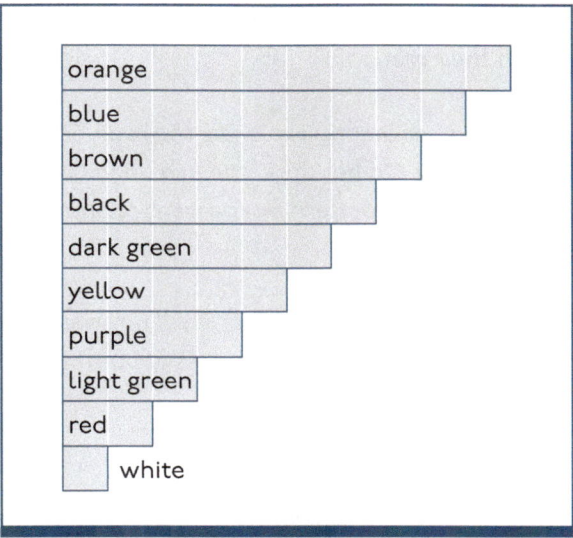

After students had the opportunity to explore with the rods, they were asked to consider which rods showed a one-half relationship.

Me: *What does it mean to have a one-half relationship?*

Juan: *It's like when one thing is one-half another thing.*

Aisha: *Yeah. Like when you put them next to each other, you can see that one is half.*

Many students nodded their heads in agreement, so I felt comfortable letting students explore the rods looking for one-half relationships. As I walked around, I could see students lining the rods up to check. I prompted students to look for other ways once they had found at least one such relationship.

I paused students and brought them back together as a group to share some of the connections they made between the rods and one half. This proved to be highly

---

[1]Examples from classrooms are based on real experiences. However, throughout we have changed the names of educators and students to protect their privacy.

accessible to students. All students were able to find at least one, and we settled on some statements that helped us think about why these relationships were one half.

- *It takes two of the yellow rods to cover the orange rod, so the yellow rod is half of the orange rod.*

- *The yellow rod covers the orange rod exactly two times with no gaps or overlaps.*

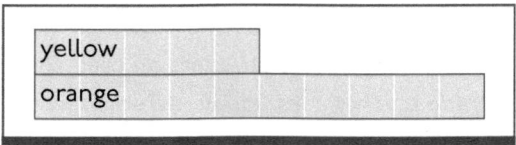

Students were then asked to find some other relationships between the rods that weren't one half. They immediately got to work grabbing rods and lining them up as they eagerly chatted with their groups.

I arrived at Lucy's desk to find her quietly exploring the rods. She looked up and asked if she could show me something. On her desk, she had lined up several different colors. There didn't seem to be any pattern or apparent relationships in the rods, but I wanted to hear what Lucy was thinking.

Lucy: *If you cut here, it goes here. Then do it again. (Lucy was holding a light green rod and making a cutting movement that was about the size of a white rod.)*

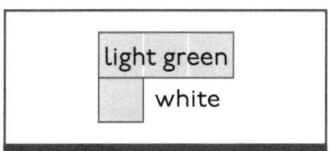

Me: *So, you would cut the green rod off here (pointing to where she indicated) and then do it again. Then what?*

Lucy: *It's three.*

Me: *So, what does that mean?*

Lucy: *It's one three.*

Me: *Yes. It would be a one-third relationship because three of those white ones fit on a light green one. Will you feel comfortable sharing that with the class when we get back together?*

Lucy: *(nods her head)*

When we returned as a group, Lucy shared her relationship first, followed by a few other students with similar findings, as well as some unique ones. Students were encouraged to work a little bit longer stretching their thinking.

As I moved around the room, I made my way back to Mrs. Yadev. I wanted to learn more about Lucy since this had been our first interaction. I noticed that her vocabulary and speech were lacking as she mostly shared through movement (head nods and arm actions). Mrs. Yadev told me that Lucy often leaves during math class to receive services for her IEP goals, and she rarely shares in class.

I wondered how often in her past Lucy had been thinking about something but wasn't able to fully articulate it. Did this prevent others from seeing Lucy as a capable math learner? When we see all of our students as capable of mathematical understanding, we start to think about ways we can help them to share their math thinking. Was it the manipulatives that gave Lucy the confidence to share out loud? Was it taking the time to listen and watch and then name the mathematical terms to connect what Lucy was saying? If we start with the assumption that all students are math learners, it changes how we provide support for students. It changes how we listen to them and seek to understand their thinking and reasoning. In her book Rethinking Disability in Mathematics, *Lambert (2024) wrote, "[W]hen we as teachers trust in the thinking of our students, especially our students with disabilities, we create the conditions our students need to trust themselves as mathematical thinkers. When we, their teachers, believe that our students can and will solve complex problems, they will be able to" (p. l.)*

Later in the class period, I returned to Lucy to check in. I wanted to see what else she was thinking about.

Lucy: *If you cut one, it almost fits. (Looking at two purple rods lined up with one black rod)*

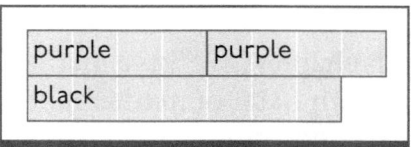

Me: *So, you're saying, if you cut this little piece off of here, the purple is almost half the black. Did I say that correctly?*

Lucy: *(Nods)*

Me: *So, you could say that you almost found another one-half relationship.*

Lucy smiled and continued working. Lucy's lack of vocabulary didn't hold her back from successfully making sense of fractional relationships. Although she hadn't found another one-half relationship, her noticing about two rods that were close to one half told me she understood what a one-half relationship should look like.

When we begin our school year and seek to learn about our students and build our classroom communities, our belief that all our students are math learners is essential. It opens the doors for our students to show us where they shine and focuses us on our students' assets.

Anchor Lessons

Kobett and Karp (2020) challenged us to view each of our students as mathematically amazing and to find and name each child's unique strengths:

> Recognizing and celebrating students' strengths through the mathematical processes they use is strengths finding at its best! As you engage students in mathematical problem solving, communicating about mathematics, representing mathematical ideas, reasoning about mathematics, and seeking and recognizing mathematical connections, celebrate the unique and brilliant ways that your students make sense of the mathematics they are learning. (Kobett & Karp, 2020, p. 60)

At the beginning of the year, teachers invest time turning their classrooms into learning communities. In a powerful math community, because as teachers we believe all students are capable math learners, we begin the year acting as "strengths detectives" (Kobett & Karp, 2020, p. 4) on a quest to get to know each student's unique personal and learning strengths. Often students' strengths are revealed as they engage in the habits of mathematical thinking.

In this chapter, we will offer a series of anchor lessons, one for each habit of mathematically powerful people. These lessons are designed to introduce students to the eight math habits that are outlined in each of the subsequent chapters and allow teachers to discover their student's mathematical brilliance.

Each anchor lesson in this chapter involves a mathematics task that engages students in the targeted math habit of mathematically powerful people. In the class discussion at the end of the lesson, students discuss their use of the math habit and reflect on its importance.

For each anchor lesson, you will create an anchor chart that lists the math habit, as well as some actionable steps students can take to practice using the habit.

You may want to introduce two or three math habits of mathematically powerful people a week across August so that students have time to practice and become comfortable with the math habits they are learning about.

Habit 1: I Expect Math to Make Sense

WHAT?

This exploration will be centered on the importance of making sense of problems and sticking with them. Students learn that teachers are most interested in their process for solving problems, not the answer; therefore we take our time to understand and try different ways to solve problems.

During the lesson, you will be creating an anchor chart with the title "I Expect Math to Make Sense" to focus the students on ways in which they can make sense of problems.

YOU NEED:

Chart paper

Markers

Word Problem

HOW?

1. Introduce a problem by reading the task out loud to students. Ask students to follow along so they can share with a partner what they heard.

    For example, There are 600 seats in the auditorium. How could the seats be arranged into sections if each section needs to have the same number?

2. Pair students to turn and talk about what they heard and what they understood about the context.

3. When students are done sharing, write these sense-making strategies on the anchor chart. Tell students that these are two strategies that they just engaged in and can help them to slow down and make sense of problems.

    a. I can take time to think about the problem

    b. I can tell someone what I read and understand about the problem

4. Select a couple of students to share what they talked about in pairs so others can listen and check for their own understanding.

    Sample responses: There is an auditorium with seats, there are 600 seats total, the seats are in sections, the sections need to be equal.

5. Students will now work on the math problem. Distribute materials and allow students time to start working on the problem in pairs, small groups, or independently.

6. After five minutes, pause and ask students to think about what they have started working on. Encourage students to share, strategically asking students with different methods to share.

7. Before returning to work, direct students' attention back to the anchor chart and add, "I can stop and check in with someone." Tell students that this is another sense-making strategy and it supports them as they work.

8. Allow students to go back to work for as much time as is needed. Circulate and encourage students to keep trying and suggest strategies that were shared by other students.

9. After the learning experience, bring the students back together to focus on the

(Continued)

(Continued)

anchor chart of Habit 1. Tell students that you will be adding a few action steps that they can use in the future.

I can make sense of a problem.

- I can take time to think about a problem

- I can tell someone what I read and understand about the problem

- I can stop and check in with someone

Source: Istock.com/gentle studio

REFLECT

One component of our classroom community is the habit of making sense of problems. Talk with students about what they did during the lesson to make sense of the problem they solved. Keep the anchor chart in a place where the action steps can be referred to in the future.

Reflect as a classroom using the following questions for discussion:

- Why might it be important to make sense of math problems before digging in to solve them?

- How does making sense of problems help you to be a mathematically powerful person?

Habit 2: I Love Challenging Math Problems

Preparing for this math experience: A key to students' success in grappling with challenging problems is that their teacher can offer scaffolds and supports without removing the rigor. To prepare for the anchor lesson, you will want to solve the problem your students will be solving. Pay attention to your thinking and struggles as you solve the problem.

WHAT?

In this experience, students will engage with a challenging math problem and discuss some important skills needed to persevere and keep working. Robert Kaplinsky's Open Middle website https://www.openmiddle.com/ has free, vetted, challenging math problems that all students can access at the same time, providing a perfect challenge as students seek to find more solution pathways. It is appropriate for all grades K–5. These problems are called "open middle" because they all start with the same beginning and end with the same answer; however, there are multiple ways to approach and solve each problem. All students have an entry point and can easily answer the problem, but the challenge is in finding the optimal answer. We invite you to visit the website and try one to get you started.

At the end of this experience, you will introduce a new anchor chart for the math habit. Title the chart "I can solve challenging math problems" and leave room to add some action steps for students to practice the habit.

YOU NEED:

Challenging Math Problem (Consider checking out the Open Middle website for problem selection.)

Chart paper

Markers

Paper

Pencil

HOW?

1. Select a problem from the Open Middle website. Problems are sorted by grade level and math strand, such as numbers and operations or geometry. You may choose to use a challenging math problem from your curriculum.

2. Tell students that they will be working on a challenging math problem today. If you are using an Open Middle problem, tell students that the challenging but fun part is that the problems have more than one pathway to arrive at a solution. They should try to solve the problem and then stop to share with a partner about what they might try next.

3. Show students the problem. Read the problem and ask students to share what they know and understand before they start working.

4. Let students work for five to seven minutes, encouraging them to share their ideas with their peers.

5. Bring the group back together to share some of the ways students are approaching the problem.

6. Students should return to working on the problem as you circulate and listen, ask questions, and offer scaffolds if needed.

(Continued)

(Continued)

7. Once students are in a place where they have tried at least one solution, bring them back together to debrief their solutions as a class.

8. After the learning experience, introduce the anchor chart for the math habit. Tell students that you will be adding a few action steps that they can use in the future.

I can solve challenging problems.

- I can get started right away even if I don't know where the end will take me
- I can stop and check in with others
- I can talk about strategies with others
- I can listen to how my peers are solving the problem
- I can keep working even if I am not sure of the solution
- I can use what I know to help me
- I know it's ok to challenge myself

REFLECT

The real-world problems students will solve in the future will be challenging. Students need to practice building the habit of perseverance in a safe classroom setting. By taking the time to brainstorm action steps they can take when they are feeling challenged, they will have essential skills for becoming problem solvers. In the classroom, solving challenging math problems is how students can apply their skills, collaborate with others, and grow their brains. Students must have multiple opportunities to experience challenges in school and reflect on their experiences.

Reflect as a classroom using the following questions for discussion.

- What are some of the feelings you had when solving this challenging problem? How did you work through them?

- What advice would you give other students who are finding their work challenging?

- How can you support yourself and your classmates when you work on challenging problems?

- Persevering is sticking to something, even when it's a challenge. What are some ways we can persevere when we do math?

Habit 3: I Learn From Math Mistakes

WHAT?

For this experience, students will think about a time they learned something new outside of the classroom. They will reflect on the journey they took including the stumbles, challenges, and mistakes they likely had along the way.

During this lesson, you will create an anchor chart and title it "I learn from math mistakes." As a class, you will add some action steps for students to use when they make a math mistake.

YOU NEED:

Chart paper

Markers

HOW?

1. Ask students to brainstorm things they have learned outside of school (e.g., riding a bike, playing a sport, playing an instrument, cooking a recipe). List the student's ideas on the board.

2. Ask students to describe a mistake they made while they were learning these new things (e.g., falling off the bike, missing a goal, playing the wrong note, burning the food). Again, list these ideas on the board.

3. Ask students to consider how these mistakes helped them learn.

4. Show photos of a couple of famous people the students will recognize (e.g., a famous athlete or a famous singer). Ask if they think these people made mistakes as they practiced. You might show this

YouTube video: https://bit.ly/3TqBCtt (Cole, 2012)

5. Talk as a class about the value of mistakes. You may have them reference the video that they viewed.

> **I can learn from math mistakes.**
>
> - I can see a mistake as helpful to learning
> - I can say, "It's ok that I made a mistake"
> - I can look at my mistakes to help me think about changes I can make to my work
> - I can celebrate my own and others' mistakes

REFLECT

Most people, adults and students alike, tend to avoid mistakes at all costs. We often have negative feelings about making mistakes; however, once we realize that mistakes are a helpful and necessary part of learning, we begin to see their usefulness. We can shift our thinking around mistakes by simply recognizing our mistakes as an essential part of the learning process. As learners become more comfortable with recognizing and celebrating their mistakes, they are in a more optimal learning zone, leading them to learn more deeply.

Reflect as a class using the following questions for discussion:

- How do mistakes help us learn?
- What are some things you can do when you make a mistake?
- How can we support each other around making mistakes?

Habit 4: I See Myself as a Mathematician

WHAT?

In this activity, students will explore their mathematical identities at this moment in time. They will answer a series of questions about their feelings and attitudes toward math and include a time capsule that they will see again at the end of the year.

YOU NEED:

Chart paper

Markers

Paper

Pencils

Envelopes

Mathematical Time Capsule Survey (Available as a downloadable resource on the companion website—**https://qrs.ly/ vqfnls2.**)

HOW?

- Tell students that they will be creating their own time capsule that they will open at the end of the year and show their mathematical growth.

- Start by having students complete the mathematical survey. Younger students may have the survey read to them while an adult scribes.

- Students should fold their papers and store them in envelopes with their names on the outside. You may even have students write, "DO NOT open until May ___."

- Tell students that they will open their time capsules at the end of the year to reflect on their growth as math learners.

Mathematical Time Capsule

Name: _____

Today's date: _____

1. What I want my teacher to know about me as a math learner:
2. My favorite math topic:
3. A time I was proud of my math work:
4. Something I get frustrated with in math:
5. What I want to learn in math this year:
6. Three things about me (not math related):
7. What I would tell my future self:
8. Here is a math problem I can solve:

Image source: Istock.com/armastas

- Finish the experience by completing an anchor chart titled "I see myself as a mathematician." Add some action steps for students to practice this habit.

I see myself as a mathematician.

- I can talk about my math strengths.
- I tell myself and others that I can do math.
- I do mathematical things such as talk about math, solve problems, look for patterns, and see math all around me.

REFLECT

Students must have time to continually explore their mathematical identities. As students learn more math habits and use them in different ways, their math identities grow and change. Educators can make time for students to explore and reflect on their mathematical growth. Leaders and teachers can also reflect on their own math identities each year they grow with students.

Reflect as a class using the following questions for discussion:

- Why might it be important to think about how you see yourself as a math learner?

- How can we grow as mathematicians this year?

Habit 5: I Talk About Math

WHAT?

This habit engages students in exploring and justifying their thinking. The focus will be on students defending their reasoning and thinking about the reasoning of others as they strive to make sense of problems. Students must also use mathematically precise vocabulary in their explanations.

During the lesson, you will create an anchor chart with the title "I can talk about math." Students will get the opportunity to try out some action steps that allow them to practice this habit.

YOU NEED:

Chart paper

Markers

Sticky notes

HOW?

In this experience, students will engage with a type of problem called "group the numbers." They will be given a set of numbers and will be asked to place them into two groups. Students should decide how they grouped their numbers and be able to accurately explain why those numbers were placed together.

1. Show students a set of numbers and tell them that they will get to make some choices about what to do with the numbers.
   Sample set of numbers: 48, 12, 19, 4, 26, 17, 23, 80

2. Students should record the numbers on a sticky note, one per sticky note. This will allow them to open sort the numbers into groups.

3. Ask students to start by working alone and considering groups they could sort their numbers into. They can place their numbers together in any way that makes sense to them.
   Examples: even and odd, divisible by 4, teen number, and other numbers

4. Once they have had a chance to decide, they will turn and tell a partner how they decided to place them in their groups.

5. Before sharing, provide students with a sentence stem that can support them in sharing.

6. *I put these numbers in this group because _____, and I placed the other numbers in this group because _____.*

7. Once pairs have had a chance to share, use the following questions for discussion:

   • How did you group your numbers?

   • Who did something similar and could share their reasoning?

   • Who did something different and can share why they grouped their numbers the way they did?

8. Finish the experience by introducing a new anchor chart for the math habit of "I can talk about math." Tell students that you will be adding action steps that can help them practice this habit.

I can talk about math.

- I can share what I think
- I can listen and think about what others said
- I can think about what is the same and what is different between my own thinking and the thinking of others
- I can use "I agree" or "I disagree" when I am speaking to others

REFLECT

The focus of the task is to encourage students to talk about their work and engage in discourse between themselves and their peers. They must speak about their choices and listen to their peers as they analyze what might be the same or different from their work.

In a classroom community, all students must speak and defend their choices as well as listen to the justification of their peers. By listening to others, everyone can learn from each other and the classroom works as a community.

Reflect as a class using the following questions for discussion:

- Why is it important that we can defend our reasoning?

- What are some ways we can show our peers that we are listening and thinking about their strategies and ideas?

- What does it sound like to disagree with a peer's idea respectfully?

Habit 6: I Represent Math in Different Ways

WHAT?

In this experience, students will be focusing on how mathematicians solve problems to help them understand the world around them. They use models to represent their problems, including writing equations, creating pictures, making a chart or table, or writing about their work.

During this lesson, you will create an anchor chart with action steps for students to model with math. Title the chart "I can represent math in different ways" and leave room to add the action steps listed in the lesson.

YOU NEED:

Grade-level math problem

Chart paper

Markers

Sample models that match the problem (charts, graphs, equations, pictures, expressions, words, etc.)

HOW?

1. Select a problem for students to look at that includes a variety of models.
   For example: Sun Lee is planting flowers in each window of her house. If she has ten flowers in each planter, how many flowers will she have planted?

2. Present the selected problem to students to read and discuss.

3. Tell students you will provide each group with a sample model for the problem they just read.

For example:

Drawing

Image source: istock.com/szakalikus

Equations

$10 + 10 + 10 + 10 + 10 + 10 = 60$ flowers in the windows

$6 \times 10 = 60$ flowers total

Window	Number of flowers planted	Total flowers
1	10	10
2	10	20
3	10	30
4	10	40
5	10	50
6	10	60

Representation:

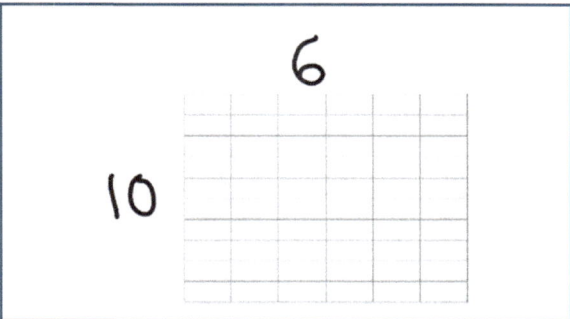

4. As a group, they should discuss how the model represents the problem and be prepared to share out. Phrases below can be written on the board to support students in looking at work:

    - *How does the model represent the problem?*

    - *How does the representation show the quantities or numbers in the problem?*

    - *How can the representation help someone to solve the problem?*

5. Ask groups to come up with a word or phrase to describe the model they are analyzing. For example, students might say that their model is a picture that shows what is happening.

6. After groups have had a chance to work, bring the whole group together to share the different models.

7. Show each model, one at a time, so all students can see it. Allow the groups to share, name, and label each model.

8. Bring student's attention to the anchor chart that you created before the lesson. Tell students that one habit of mathematicians is to make models like the ones they worked with in the lesson. You will list some action steps that students can take to practice this math habit.

I can represent math in different ways.

- I can write an equation to represent the problem

- I can create a table, chart, or graph to represent the problem

- I can draw a picture or model to represent the problem

- I can speak about the problem and write down my strategy and solution in words

- I can use tools to represent the problem

REFLECT

Another habit that mathematicians use is to model with mathematics as they make sense of problems in the world around them. Models support mathematicians in clarifying their thinking and reasoning, and they represent mathematics in a variety of ways to uncover their meaning.

Reflect as a classroom using the following questions for discussion:

- How does a model help you understand the math you are learning?

- How can a model help to communicate to others what you are thinking?

Habit 7: I Make Math Connections

WHAT?

In this lesson, students will explore math tools to help them think about the importance of making connections. Prepare an anchor chart titled "I can make math connections" to be used during the conclusion of the lesson.

YOU NEED:

Chart paper

Markers

Scissors

Tape

Five-frame (optional)

Ten-frames* (one for each student)

Double ten-frame (optional)

Hundreds chart

Number line (optional)

HOW?

1. Show students a ten-frame and ask them to share what they notice or what they know about it.

2. Display a hundreds chart and ask students to think about what is the same and what is different from a ten-frame. Listen to responses and record them on the board. Ask students the following questions:

   a. What mathematical ideas do these two tools represent?

   b. How are these tools used to do math?

3. Tell students that they are going to show how a ten-frame is connected to a hundreds chart. Give each student a paper ten-frame. Ask students how they could modify the ten-frame so it looks like a group of ten on the hundreds chart. Students will likely suggest they can cut the ten-frame so they have two groups of five. By placing the two groups next to each other in a line, they now have a row of ten.

4. Students should cut their ten-frames, rearrange them, and tape them back together to make long rows of ten boxes.

5. Discuss with students how they could lay a modified ten-frame over a row in the hundreds chart. Tape the new ten-frame over the first row of boxes on the hundreds chart. Continue adding rows from students' modified ten-frames. Each time, confirm that students still have ten boxes.

6. Ask students to think about other math tools that are connected to the ten-frame and/or hundreds chart. Students should share their reasoning.

7. You might also explore the relationship between five-frames and ten-frames, ten-frames to double ten-frames, or the hundreds-chart to a number line.

8. After the experience, show students the anchor chart titled "I can make math connections." List some action steps that students can take to make connections, notice connections, and use connections to learn math.

I can make math connections.

- I can notice how things are alike
- I can think about how my problem, strategy, and solution is related to someone else's
- I can look for similarities and differences between new problems and ones I have solved
- I can think about connections between math ideas

REFLECT

Learning math deeply requires students to make connections between math concepts, problems, and mathematics to themselves. Educators can support students in developing a deep understanding of math content when they encourage students to connect their learning.

Reflect as a class using the following discussion questions:

- How can looking for connections support us in learning math?
- What connections can you look for?

*Note: Make sure the ten-frame boxes and the hundreds chart boxes are the same size.

Habit 8: I Look for Patterns

WHAT?

In this experience, students will learn how mathematicians look for patterns and structures to make sense of the world around them. Taking the time to notice patterns will help students to deconstruct complicated problems into their parts. Students will use the patterns and structures to connect their learning from what they know to what is new.

At the end of this experience, you will introduce a new anchor chart for the math habit. Title the chart "I can look for and use patterns to help me in math" and leave room to add some action steps for students to practice the habit.

YOU NEED:

Chart paper

Markers

String of math problems that are grade level appropriate

HOW?

- Gather a series of related problems (six to eight problems is recommended).

  For example:

  1 × 4

  2 × 4

  4 × 4

  8 × 4

  8 × 8

- Tell students that they will be looking for patterns. During the task, everyone will remain silent so they can focus on patterns they see.

- Go over some guidelines for silent math.
  - A sticky note will be placed on the board, which means everyone is silent (including the teacher).
  - The teacher will write a problem and ask for a volunteer to come up and write the answer.

    For example 1 × 4
  - Students will give a thumbs up if they agree or thumbs down if they do not.
  - If students agree, continue writing equations until all equations have solutions written next to them.

    For example:

    2 × 4 = 8

    4 × 4 = 16

    8 × 4 = 32

    8 × 8 = 64
  - If students disagree, allow the student who wrote the answer to change their answer or select a peer to change the answer.
  - After all equations have been written and solved, remove the sticky note.
  - The class will now discuss the problems.

- Engage students in a discussion about what they notice about the numbers. Sample discussion prompts include:
  - *What patterns do you notice in the problems?*
  - *What patterns do you notice in the solutions?*
  - *Why do you think that happens?*

    For example:

    When the first factor doubles and the second stays the same, the product doubles.

- Once students have had a chance to engage in silent math and discuss their connections, bring them back together to discuss the math habit.

- Tell students that another habit that mathematicians use is to look for patterns. Show students the anchor chart that you prepared, and record some action steps that students can use when practicing this habit.

I can look for and use patterns to help me in math.

- I can look at math problems and think about what I notice

- I can seek out patterns often

- I can create patterns

- I can ask myself, "Is there something that is repeated?"

- I can think about what keeps happening with other related problems

---

REFLECT

Mathematicians build off of what they know by seeking structure in new problems. Taking the time to look for patterns allows them to make connections between problems. When mathematicians use what they know, they are presented with solution paths for solving new problems.

Reflect as a class using the following questions for discussion:

- How might looking for a pattern help you?

- Where can you see patterns in math?

LET'S DO SOME MATH ACROSS OUR SCHOOL!

In building a powerful math community made of individuals who see themselves and others as mathematicians, students need ample experience exploring the math around them. The tasks we select serve as a mirror reflecting the mathematical richness in each of our students.

Tell students that they will get the opportunity to consider numbers that reflect something about them. Students should consider the question, "How could I quantify something about me?"

1. Write 10–15 numbers on the board that are significant to you. *For example, you might write numbers such as 6, 1985, 0.19, and 38.*

2. Give a clue about each number. *For example, "One of these numbers represents the year I was born."*

3. Ask students to decide which number matches the clue they were given.

4. Repeat the process for each number.

5. Next, have students write down 10–15 numbers that are significant to them.

6. On a separate piece of paper, students will write a sentence that describes each of their numbers.

7. Students will work in pairs to exchange their number lists and take turns reading the clues. Students should read their clues randomly while their partner tries to guess the number.

For students in younger grades, this activity may be done orally. You may have students think of a number that represents something about themselves. It may be helpful to provide your students with some examples such as the number of pets or siblings they have.

LITERATURE CONNECTIONS

***Look: I'm a Mathematician* by Hilton** (2019) (Grades K–2))

This book begins by telling readers that they use their brains and senses to do mathematics. It then offers 14 STEM-like activities representing a variety of mathematical strands and topics to engage children with mathematics.

Activity: How do you use each of these types of mathematics inside and outside of school: numbers, patterns, shapes, and measurement? Make lists of ways you use each type of math. Add to your lists as you think of more ideas.

***Ada Byron Lovelace and the Thinking Machine* by Wallmark** (2015) (Grades 3–5)

This picture book biography tells the story of Ada Lovelace, a renowned 19th-century female mathematician and the first computer programmer. Ada's story shows how the passion she felt for numbers and mathematical problems as a child grew into her vision of a computing machine that, as Ada correctly predicted, changed the world in amazing ways.

Activity: Ada Lovelace is famous because she used mathematics to solve real-world problems and make the world a better place. What is a real-world problem that you care about? How might mathematics be used to help solve that problem?

SPOTLIGHT ON BRAIN SCIENCE

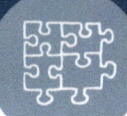

Our students need to know that human beings are built to learn. We can't help it! Our brains are designed to use our daily experiences to construct new understandings and develop new skills. When we accept that we are learners by nature and when we consider how learning happens in our brains, we can learn more effectively and efficiently (Oakley et al., 2021).

You can share the following information with students in a mini lesson or a station activity. You might also send this reproducible home along with the Family Newsletter to help families talk about these important ideas.

WE LEARN WHEN OUR NEURONS COMMUNICATE.

The study of brain science is called "neuroscience," and scientists who study how the human brain works are called "neuroscientists." You can and should be a neuroscientist today and every day of your life because when you understand how your brain works, you can learn more and do more of the things that are important to you.

Our brains are always changing. Every experience we have in and out of school changes our brains and helps us learn. This is a picture of a brain cell, called a "neuron."

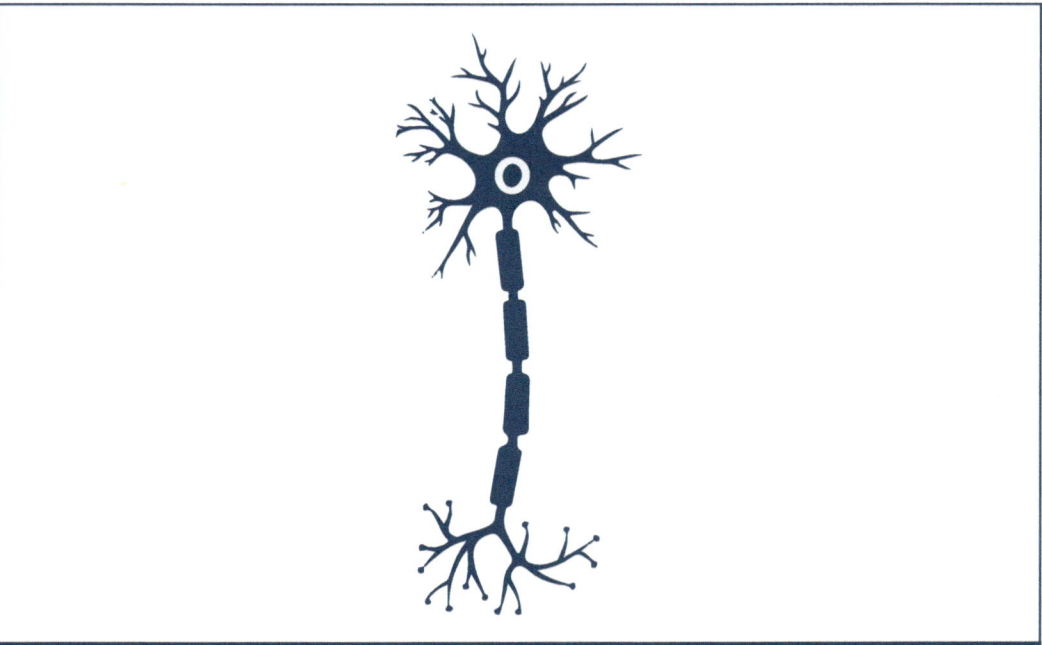

Image source: Istock.com/drypsiak

When you think about a math problem, your neurons communicate with each other through chemical signals. Every time you think about math problems, these communication pathways become stronger. You also grow new pathways. That means you're learning!

Choose one of these ways to think more about this important scientific information:

1. Write down why these ideas are important to you and how you want to put them to use in your life.

2. Talk to a friend about these ideas. Decide together how you will put them to work.

3. Share these ideas with someone at home. Tell them why these ideas are important to you and how you plan to use them.

 Available as a downloadable resource on the companion website.

SPOTLIGHT ON EQUITY: POSITION ALL STUDENTS AS VALUED MATH THINKERS

It is a terrible truth that students who are different from their classmates are, at times, given a lower status within the class community. Marginalized students may include those who receive math instruction outside the core classroom, as well as students who have differences related to race, religion, and native language. Peers sometimes assume, often unconsciously, that these differences indicate a classmate is less capable or less important. In a powerful math community, teachers actively watch for indications of negative bias toward any student and take firm steps to address and reverse this bias.

When we as teachers believe that all students are capable math learners, we look for and find each student's mathematical brilliance. We can raise the status of marginalized students by revealing and calling attention to their strengths (Boaler, 2015).

Picha (2022), author of *Conferring in the Math Classroom,* uses one-to-one conferences to help students learn to see themselves as mathematically capable and to help others notice these students' mathematical brilliance. Picha's math conference structure includes three essential elements:

- **Listen and Observe**—As the teacher sits beside a student, they watch and listen for evidence of math habits and math understandings, paying attention not only to the student's words but also to their nonverbal communication.

- **Name the Strength**—Of the competencies noted, the teacher selects a specific strength to name out for the student with the goal of building the student's awareness of and pride in this mathematical asset. The teacher describes the strength in language the student can understand, provides evidence of this strength, and briefly explains the importance of the strength.

- **Encourage Students to Share Their Thinking**—The teacher supports the student in sharing their math thinking with a partner, a small group, or in a whole-class discussion. This sharing provides the student with practice communicating about their math thinking while strengthening the student's math agency and identity. Most important, it publicly assigns competence to the student in front of peers.

Picha believes these five-minute math conferences can make a critical difference in a child's life. In her book, she stated, "I am hard-pressed to think of anything more rewarding than watching students discover the world of mathematics isn't something they need to earn their way into. They already belong in this world, and sometimes we get the honor of watching them walk in" (2022, p. 21).

TRY IT

Select a student who might not share often in class. Set aside a five-minute window during your week where you can sit side by side with the student and really listen to their thinking and observe them work. Prepare some questions that model curiosity into student thinking.

- How are you thinking about this?

- What makes sense to you?

- What parts do you know? Which ones are you unsure of?

When the student is ready, make time for them to share their ideas with the class. With your students, start a discussion of what it means to see everyone as a math learner. Guide the discussion using the following questions:

- How can we see ourselves as math learners?

- How can we see others as capable math learners?

MATHEMATICAL ME: STUDENT JOURNAL AND PORTFOLIO

Student Journals:

Have students respond to the following question:

- How do I know I am a math learner?

Older students can write their responses in their Mathematical Me journals. Younger students can draw a picture showing their response, or you might gather responses during a class discussion or quick one-on-one interviews.

Review students' responses to monitor students' beliefs about learning mathematics. You might also tally or graph the different ideas that students offer.

Be sure to share and discuss this data with your class. Ask, "What does this data tell us?" and "What are some things we can do to continue growing as mathematicians?"

Student Portfolios:

Distribute the Math Attitude Survey to students at the beginning of the year. Let them know that you are interested in learning how they feel about themselves as math learners.

Grades K-1

Name: _____ Date Taken: _____

Math Attitude Survey

Grade K–1

Direction for Teachers:

1. Read each question to students, one at a time.
2. Ask students to circle the faces that most closely matches how they feel about the question.
3. Clarify any questions as needed.

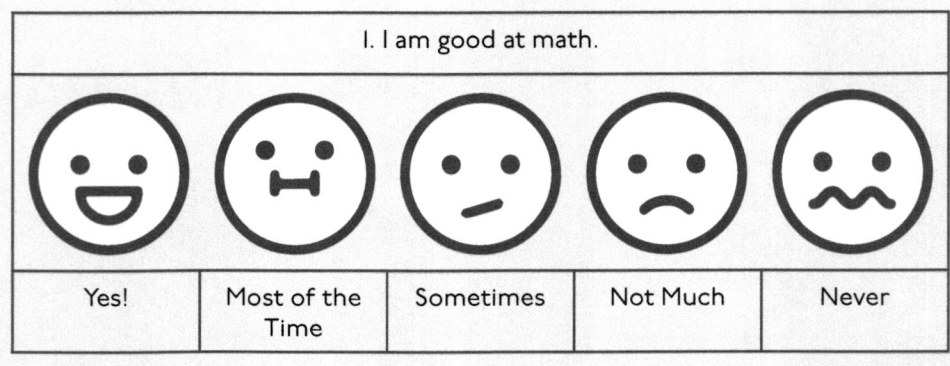

1. I am good at math.				
Yes!	Most of the Time	Sometimes	Not Much	Never

Images source: istock.com/calvindexter

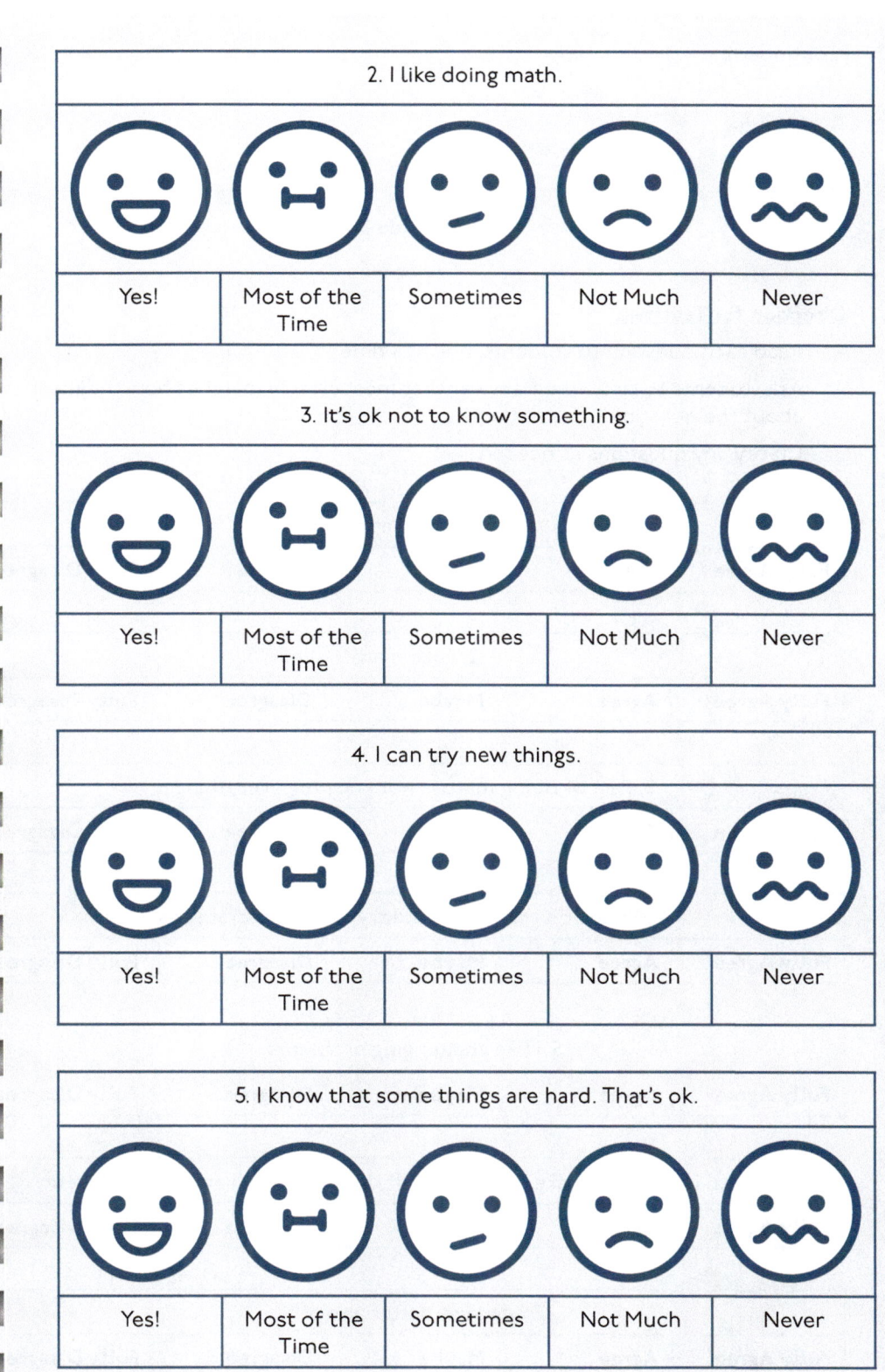

Images source: istock.com/calvindexter

online resources — Available as a downloadable resource on the companion website.

(Continued)

(Continued)

Grades 2-5

Name: _____   Date Taken: _____

Math Attitude Survey

Grade 2–5

Direction for Teachers:

1. Read each question to students, one at a time.
2. Ask students to circle their answer that most closely matches how they feel about the question.
3. Clarify any questions as needed.

1. I am good at math.				
Fully Agree	**Agree**	**Maybe**	**Disagree**	**Fully Disagree**

2. I like math.				
Fully Agree	**Agree**	**Maybe**	**Disagree**	**Fully Disagree**

3. Part of doing math is not knowing everything.				
Fully Agree	**Agree**	**Maybe**	**Disagree**	**Fully Disagree**

4. When I do math, I should try different strategies.				
Fully Agree	**Agree**	**Maybe**	**Disagree**	**Fully Disagree**

5. I like challenging problems.				
Fully Agree	**Agree**	**Maybe**	**Disagree**	**Fully Disagree**

6. When I make a mistake, it's normal. It doesn't mean I am not intelligent.				
Fully Agree	**Agree**	**Maybe**	**Disagree**	**Fully Disagree**

7. I can get better at math.				
Fully Agree	**Agree**	**Maybe**	**Disagree**	**Fully Disagree**

online resources   Available as a downloadable resource on the companion website.

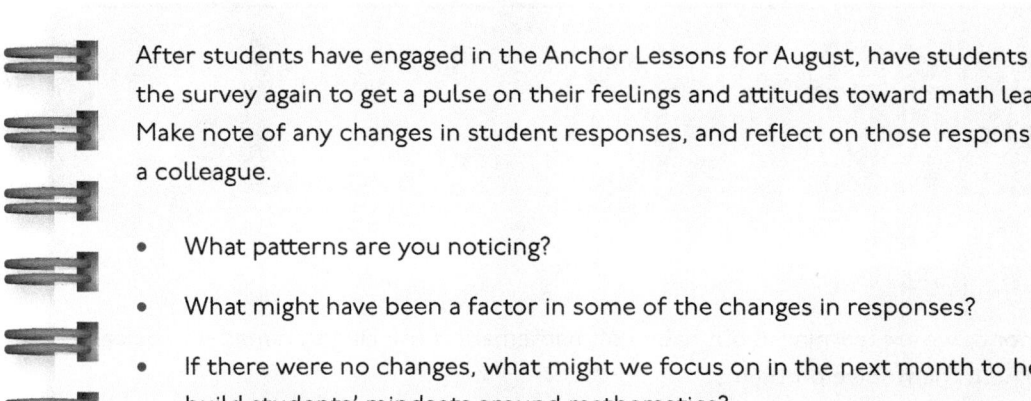

After students have engaged in the Anchor Lessons for August, have students take the survey again to get a pulse on their feelings and attitudes toward math learning. Make note of any changes in student responses, and reflect on those responses with a colleague.

- What patterns are you noticing?

- What might have been a factor in some of the changes in responses?

- If there were no changes, what might we focus on in the next month to help build students' mindsets around mathematics?

Save completed surveys in students' math portfolios. Students will retake this same survey in December and May and reflect on their growth.

FAMILY NEWSLETTER: WE ARE ALL MATH LEARNERS!

WHAT WE'RE LEARNING

This month we are learning about habits of mathematical thinking that mathematicians use to help them solve problems.

WHY IT'S IMPORTANT

People sometimes think that math only involves adding, subtracting, multiplying, and dividing numbers to get correct answers. But math is so much more than just calculation. Mathematicians use specific habits of thinking, including making sense, taking on challenges, reasoning and communicating their thinking, learning from mistakes, making connections, seeking out patterns, and reflecting on their own attitudes and beliefs to help them solve problems. Your child will have the opportunity to strengthen these habits of mathematical thinking across the school year.

HOW YOU CAN HELP

Ask your child about the math habits that they are learning about and using in school.

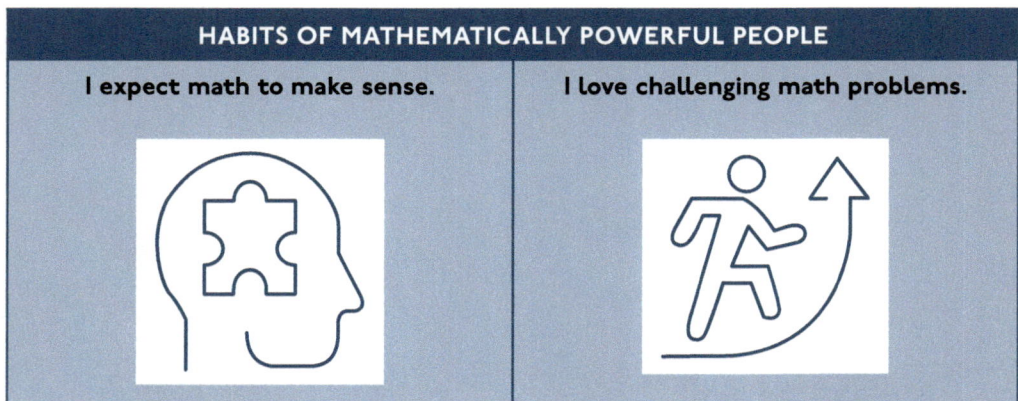

Images source: istock.com/Fourleaflover

Images source: istock.com/Fourleaflover

When your child works on math at home, you can ask the following questions to support their growth.

(Continued)

(Continued)

I expect math to make sense	Does that make sense to you? Why or why not?
I love challenging math problems	What math ideas could help you to figure this out?
	I know this is challenging. What can you try?
I learn from math mistakes	What part of this answer is correct? What can you try now since you know what the answer isn't?
I see myself as a mathematician	What math strengths do you have? Tell me about them.
	How can you use your strengths to help you?
I talk about math	Tell me about your math thinking.
	Tell me how you figured that out.
I represent math in different ways	Could you draw a picture? Can you use numbers or words to describe what you are thinking?
I make math connections	What parts do you know? Have you solved problems like this before?
I look for and use patterns	What are you noticing?
	Have you seen a similar problem before?

online
resources ☐ Available as a downloadable resource on the companion website.

CHECKING IN ON OUR LEARNING: WE ARE ALL MATH LEARNERS

- How do I know I am a math learner?

This month's focus, *we are all math learners,* helps us to see all of our students as capable of learning mathematics. We reflect on how math is used in our own lives and consider the habits that we use in doing and learning math. By introducing our students to the habits of mathematicians, our students begin to see what they do and how they think in math class as relevant and necessary to recognizing themselves as math learners.

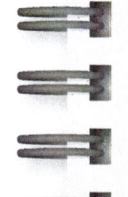

MATHEMATICAL ME: EDUCATOR JOURNAL

- How has this month's learning focus supported your students' mathematical growth?

- How has this month's learning focus supported your growth as a math teacher?

- How has this month's learning focus supported your school's growth as a powerful math community?

LOOKING AHEAD

In September, we will explore the role of sense-making in mathematics. Leaders, teachers, and students will consider the shift from seeing math as facts to be memorized to a subject that can and should be understood.

Questions for Reflection

- What are some ways you make sense of any type of problem you encounter?

- How do you help students make sense of challenging problems?

CHAPTER THREE

September

We Expect Math to Make Sense!

ESSENTIAL QUESTIONS

- Do I expect math to make sense?

- What should I do when a math idea doesn't make sense to me?

THIS MONTH'S FOCUS

In a powerful math community, students and educators expect math to make sense and know they are capable of making sense of mathematical ideas. Cognitive dissonance is recognized as a natural part of the sense-making and learning process.

Educators help students to see themselves as capable of learning math with understanding. They take responsibility for developing students' mathematical agency, as well as for their mathematical proficiencies.

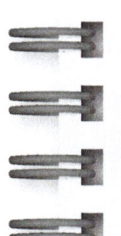

MATHEMATICAL ME: EDUCATOR JOURNAL

Take five minutes to respond to the essential questions. You will have a chance to reflect on these same questions and to have students respond to these questions at the end of the month.

ON YOUR OWN

LET'S DO SOME MATH! FIST BUMP PROBLEM

One morning before school, you and the other teachers on your grade-level team gather informally in the hallway to check in with each other. The conversation turns to students' reactions to the math habits anchor lessons you taught in August. Team members trade stories about students' excitement and insights during these lessons, and you realize your students are genuinely proud to see themselves as mathematicians. Inspired by the students' enthusiasm, your team considers how this month's focus on mathematical sense-making will continue strengthening students' mathematical identities and sense of community. You suddenly notice that it's almost time for students to arrive. As you and your colleagues hurry off to classrooms, you quickly celebrate the math learning that you are seeing with a joyful round of fist bumps.

Image source: lstock.com/leremy

Just for fun, let's add a little math to this context.

If each teacher on your team gives every other teacher on your team a fist bump, how many fist bumps will there be?

Solve this problem in a way that makes sense to you. Record your math process on a piece of paper. Save this record of your thinking to share with colleagues in Together With Your Learning Community. Consider how you engaged in the habits of mathematically powerful people as you thought through and solved the problem.

Problem inspired by the Fist Bump Problem in *Engaging in Culturally Relevant Math Tasks: Fostering Hope in the Elementary Classroom* (Matthews et al., 2022, p. 27)

 Available as a downloadable resource on the companion website.

Why This Focus

Sue's Math Story:

Cooking offers many mathematical sense-making opportunities. Although I don't consciously think about it, my cooking projects often involve a series of mini math problems.

One day, I made a half recipe for my family's favorite chocolate chip cookies. Here is the list of ingredients needed for a full recipe:

- 2 ¼ cups flour
- 1 ¼ teaspoons salt
- 1 teaspoon baking soda
- 1 cup butter, softened
- 1 cup white sugar
- ½ cup brown sugar
- 1 egg
- 2 tablespoons milk
- 1 ¼ teaspoons vanilla extract
- 2 cups semisweet chocolate chips
- ¾ cup chopped pecans

I took out a mixing bowl, measuring tools, and the canister of flour. "Half of two-and-a-fourth cups is one-and-an-eighth cups. My smallest measuring cup is one-fourth cup. How can I measure one-eighth cup? Hmmm … I can figure this out. There are 16 tablespoons in a cup, so each tablespoon is one sixteenth of a cup. So, two tablespoons would be one eighth of a cup."

I measured out the next several ingredients without difficulty but had to pause again to think about the needed quantity of pecans. "What's half of three fourths? Three eighths. Is that right? I'm splitting this recipe in half. It feels like I should be dividing. Hmmm … Oh, I've got it now—When I multiply a quantity by two, I am doubling that quantity. When I multiply a quantity by one half, I am halving that quantity. Dividing by two is the same as multiplying by one half. But three eighths is almost one-half cup. Everyone loves nuts! I'll just go with one-half cup."

The recipe called for baking the cookies at 375° for 12 minutes. But I wanted to use my new air fryer. The directions that came with my air fryer said I should cut the conventional baking temperature by 25° and reduce the cooking time by 20%. "So, I'll set the air fryer for 350°. What about the cooking time? Hmmm … I can think this through. If I cut the time by two minutes, that's one sixth of the original 12 minutes. One third is about 33%, so one sixth must be about 16%, a little less than 20%. So, I think the cookies need to bake for a little less than 10 minutes." I pulled out the calculator on my phone to check my thinking, entering 12 × 0.8, coming up with 9.6. I set the timer on the air fryer for nine minutes and 30 seconds.

The cookies, by the way, were delicious!

> "Mathematics is one of the finest tools ever conceived by the human mind for making sense of the world around us."
>
> —Woo
> (2019, p. 274)

We engage in mathematical sense-making all the time in daily life. We know that math is rooted in logic and that we can count on it to behave in predictable ways. The expectation that *math makes sense* and the belief that *we can make sense of math* are foundational to mathematical thinking.

And yet, we all know students who do not expect math to make sense and students who do not trust that they are capable of making sense of mathematical ideas. When students give unreasonable answers to computational problems and don't recognize these answers as unreasonable, they're showing us that they don't expect math to make sense. And when students can't get started on problems or guess at answers, they're showing us that they don't believe in their ability to make sense of mathematics.

Cathy Seeley, former President of the National Council of Teachers of Mathematics, stated:

> Students need to develop a habit of expecting, insisting, and demanding that mathematics make sense, whether it's an answer to a problem they're solving, an explanation someone else is presenting about their thinking, or a new mathematical idea they're learning. When something doesn't make sense in mathematics – when what they just did or found or heard or read isn't reasonable – they need to pay attention. Maybe they need to ask a question for clarification. Or maybe they need to reconsider the approach they just tried. Regardless, their cognitive dissonance – the realization that

something doesn't make sense – should drive them to dig deeper until whatever it is does make sense. (2015, p. 303)

Cognitive Dissonance

James Nottingham, author of *The Learning Challenge* (2017), calls the experience of **cognitive dissonance** "being in the pit" (see Figure 3.1). "The pit" represents "the mental discomfort produced when one is confronted with new information that contradicts prior beliefs and ideas" (p. xxiv).

Figure 3.1 • Cognitive Dissonance (aka "The Pit")

Image source: Istock.com/PhotonPhotos

Nottingham said:

> When your students get into the pit, you should expect them to feel uncomfortable. I don't mean anxious. I don't mean overwrought or afraid. I mean the opposite of contented. I mean needled; spurred on to think more, try more and question more. (2017, p. 16)

Boaler told us that when we experience the discomfort of cognitive dissonance, important learning is actively occurring. She shared this story of a student and a teacher's dialogue about this precise moment:

Student: Ms. Schaefer, I am really in the pit!
Teacher: Excellent! What classroom tools do you need? (2019, pp. 65–66)

Making sense of complex mathematical ideas requires that we are okay with cognitive dissonance and recognize being "in the pit" as part of learning.

Once we realize we are experiencing cognitive dissonance, we need to activate our beliefs that math is supposed to make sense and that we are capable of making sense of math so that we can climb out of "the pit."

Seeley (2015) described the internal process that drives learners to climb out of cognitive dissonance. She said we can help students develop an "inner mathematical voice" that shouts out when something doesn't make sense: "Wait a minute. That doesn't make sense. And math is *supposed* to make sense!" (p. 303).

> Making sense of complex mathematical ideas requires that we are okay with cognitive dissonance and recognize being "in the pit" as part of learning.

Image source: istock.com/Mooikunst

Mathematical Agency

This insistence that math is supposed to make sense is related to the idea of **mathematical agency** (National Council of Teachers of Mathematics [NCTM], 2020). When learners have agency, they initiate action to support their own learning; they are active agents in their learning rather than passive recipients. Students with agency engage enthusiastically with mathematical problems, relishing the chance to grapple with challenging tasks. They ask questions, share their math thinking, and make use of mathematical tools. They expect math to make sense and see themselves as mathematical sense-makers. They take "ownership of mathematical meaning" (p. 51).

> "Making space for students to develop agency is in and of itself an act of love."
>
> —Watson (2022, p. 517)

Tapping Into Your Experience

Metacognition

Seeley's (2015) inner mathematical voice actually references the process of **metacognition**, the ability to think about our own thinking. This is a kind of self-talk. We can help students strengthen this inner mathematical voice by helping them learn the skills of self-questioning and self-reflection.

Metacognitive Questions to Support Sense-Making

- Is this answer reasonable?

- Does this idea make sense to me?

- How does this idea connect to what I already know?

- Does this idea make me question what I thought I knew?

Consider:

- Why are these metacognitive questions powerful?

- When do you ask yourself questions like these?

- Which of your students ask themselves questions like these? Which students could use support in learning to ask themselves these metacognitive questions?

- How might you help your students build the habit of asking themselves sense-making questions? How could this habit impact their learning and view of themselves as mathematical beings?

Research has shown that metacognition boosts learning. *Visible Learning* researchers Hattie et al. (2017, p. 185) reported an effect size of 0.69 for instructional strategies that support metacognition. (Instructional strategies with effect sizes above 0.40 are considered highly effective.)

Strategic Questioning to Promote Sense-Making

The metacognitive questions students learn to ask themselves are often a reflection of the questions they are asked by teachers and others. The act of teaching involves asking lots of questions. We use questions to assess understanding, to spark thinking, and to focus students' attention on important mathematical ideas. In *Taking Action,* Huinker and Bill (2017) explained, "Through purposeful questioning, teachers send important messages to students that learning mathematics includes reasoning quantitatively and abstractly, constructing explanations and justifications, examining the reasoning of others, and making sense of mathematics" (p. 115).

In the *5 Practices for Orchestrating Productive Mathematics Discussions,* Margaret Smith and Mary Kay Stein (2018) differentiated between **assessing questions** designed to surface student thinking and **advancing questions** intended to extend learning. Consider the purposes of these two types of questions in Table 3.1. Then, look at the map for how these two types of questions might be used to encourage sense-making (Figure 3.2).

Table 3.1 • **Purposes for Assessing and Advancing Questions**

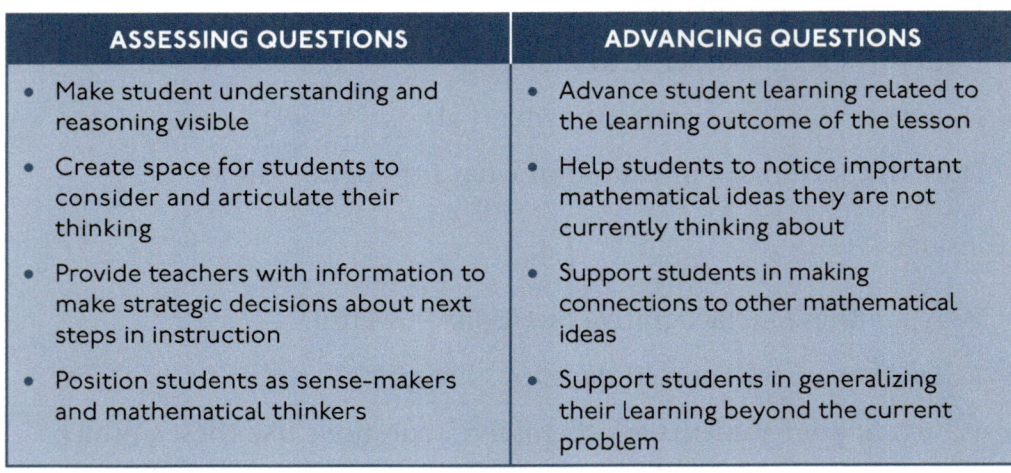

ASSESSING QUESTIONS	ADVANCING QUESTIONS
• Make student understanding and reasoning visible • Create space for students to consider and articulate their thinking • Provide teachers with information to make strategic decisions about next steps in instruction • Position students as sense-makers and mathematical thinkers	• Advance student learning related to the learning outcome of the lesson • Help students to notice important mathematical ideas they are not currently thinking about • Support students in making connections to other mathematical ideas • Support students in generalizing their learning beyond the current problem

Source: Adapted from Smith & Stein, 2018, p. 44.

Figure 3.2 • **Map for Using Assessing and Advancing Questions to Promote Sense-Making**

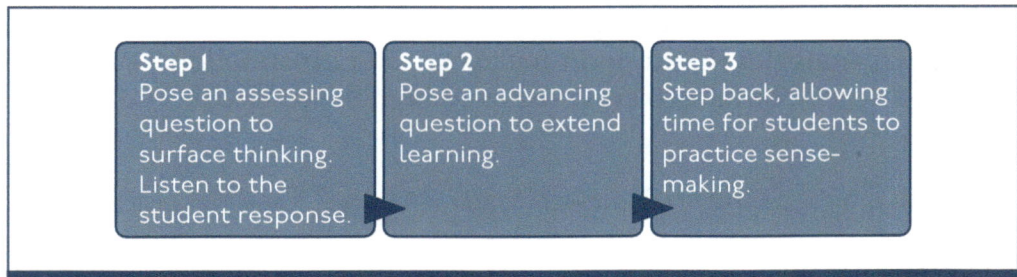

Step 1
Pose an assessing question to surface thinking. Listen to the student response.

Step 2
Pose an advancing question to extend learning.

Step 3
Step back, allowing time for students to practice sense-making.

Source: Adapted from Smith & Stein, 2018, p. 44.

Here's what this questioning sequence might look like in the context of problems from this chapter.

The Fist Bump Problem

1. Assessing question: What is your strategy for keeping track of the fist bumps?

2. Advancing question: How might you figure out how many fist bumps there would be for any number of people?

3. Step back, allowing time for students to practice sense-making.

The Making Cookies Problem

1. Assessing question: How might you represent three fourths divided by two with a picture?

2. Advancing question: How is division of a fractional number similar to and different from division of a whole number?

3. Step back, allowing time for students to practice sense-making.

Give it a try. Think of a challenging mathematics problem you will use with students the next day or so. Write down an assessing question you will ask to uncover student understandings related to the problem. Write an advancing question you will ask to help students generalize beyond the problem or apply their learning to a new context. Try out your questions. After asking your advancing question, step back and observe what happens.

TOGETHER WITH YOUR TEACHING COMMUNITY

Since We Met Last (10 minutes)

In August, you and your colleagues collected some classroom data about students' current proficiencies with the math habits.

Use the following protocol to share and reflect on this data with your math teaching community:

- Share your data with a colleague. (5 minutes)

- As a group, discuss what you learned from the data and how these insights can help you to support your students' growth as mathematicians. (5 minutes)

Let's Do Some Math Together! (10 minutes)

In small groups, share your solutions to the Fist Bump Problem. Look at and appreciate all the different ways that team members thought about and represented the problem. Enjoy a celebratory round of fist bumps if you wish.

Consider how many fist bumps there would be if one, two, or three more teachers joined your group. What if the entire faculty was together? Is there a way of determining the number of fist bumps for any size group?

Discuss:

- What math habits did you and your colleagues engage in as you solved the problem and discussed your solution methods?

- How did you experience cognitive dissonance as you made sense of the problem? What helped you to make sense of the problem?

- How did you use the metacognitive skills of self-questioning and self-reflection as you made sense of and solved the problem?

- What are you noticing about your mathematical agency as you work on and talk about challenging mathematics problems?

Building Our Expertise (30 minutes)

The NCTM has stated that students need regular opportunities to practice mathematical sense-making. In their landmark publication *Catalyzing Change in Early Childhood and Elementary Mathematics*, the NCTM stated, "Children deserve mathematically powerful learning spaces that emphasize reasoning and sense making on a daily basis with every mathematical encounter" (2020, p. 77).

Your school or team can promote mathematical sense-making by creating opportunities for students to practice sense-making within the context of your problem-solving lessons.

Together with your colleagues, use these steps to plan a problem-solving lesson that allows your students to practice the skills of mathematical sense-making:

1. Select a thinking-rich math problem to use as the basis of a problem-solving lesson.

2. Identify your mathematics learning goals for this lesson.

3. Try out the problem yourself, ideally with colleagues. Note opportunities for mathematical sense-making in the problem-solving process.

4. Plan assessing and advancing questions aligned with the lesson's learning goals and ways you anticipate students may approach and solve the problem.

5. Teach the lesson. As you pose advancing questions, be sure to walk away to allow students to practice mathematical sense-making. Notice what happens.

What are you learning about the teacher's role in helping students to see themselves as mathematical sense-makers?

How might you incorporate these planning-for-sense-making steps into your weekly lesson-planning routines?

Let's Try It (10 minutes)

Sense-making is an internal process and, as such, cannot be directly measured. We can, however, gather important data related to our students' sense-making processes to inform our decisions about how to best support students as math learners. As a teaching community, choose one of the data-collection focuses below and record this task and deadline in your calendar or planner. Bring this classroom data to next month's meeting to share and discuss.

> **Option 1: Do students expect math to make sense?** Across the month, collect brief anecdotal notes about individual students' words and actions that indicate whether or not they expect math to make sense. Do they recognize unreasonable answers to problems as being unreasonable? Do they persevere with challenging problems or give up?

(Continued)

(Continued)

Option 2: Can students identify moments when math doesn't make sense? An important step in sense-making is recognizing when something doesn't make sense. Tell students that confusion is a natural part of the learning process. You might introduce the terms "cognitive dissonance" or "being in the pit" to give students a label for the experience of feeling confused. Encourage students to verbally identify points when they are confused, or to reflect on the experience of being confused in their math journals or exit cards. Tally the number of times students can identify this point on a daily or weekly basis. Are students becoming more self-aware of their confusion? Is this helping your class to talk about ways of pushing through confusion? Are these discussions helping students to see that confusion is an expected and needed part of learning?

Option 3: How do advancing questions support sense-making? Think of a challenging mathematics problem you will be using with students in the next day or so. Write down an assessing question you will ask to uncover student understandings related to the problem. Write an advancing question you will ask to help students generalize beyond the problem or apply their learning to a new context. Try out your questions. After asking your advancing question, walk away and observe what happens. Take some anecdotal notes to share with colleagues.

Take time now to develop your plan for collecting this data. Create a simple data collection tool (tally chart or space for anecdotal notes), and place this tool in a spot where you'll be sure to use it.

ADDITIONAL PROFESSIONAL LEARNING EXPERIENCES

This month's additional professional learning experiences relate to two aspects of mathematical sense-making: **number sense** and **operation sense**. Number sense is the ability to reason flexibly about numbers and number relationships. It can be thought of as "common sense" with numbers (Burns, 2022, p. 51). Operation sense is an understanding of the meanings and interpretations of the four number operations: addition, subtraction, multiplication, and division. It involves understanding the actions associated with each operation, when it is appropriate to use each operation, and the relationships between the different operations (Van de Walle et al., 2019).

Math Talks: Tell Me All You Can

This month's math talk is a routine that can be used flexibly with any math content to build students' (and educators') number sense.

1. Write a computational problem on the board.

   Hint: Write the problem horizontally rather than vertically to encourage the group to think flexibly about different ways to solve the problem rather than defaulting to a standard algorithm.

2. Ask the group to think about all the things they can say about the problem *without giving the answer*. After a generous amount of think time, ask group members to share out and discuss ideas.

   Sentence frames such as these can be used to support thinking and sharing:

   - The answer is going to be close to/about ____ because____.

   - The answer is going to be less than/greater than ____ because ____.

Together as a math teaching community, give this routine a try. You can use a computational problem similar to problems you are currently using with students, or you might use this problem:

   25% off of $72.45

Discuss the experience:

- How did the Tell Me All You Can routine nudge you to make sense of the problem?

- How did it tap into and extend your number sense, your understanding of numbers and number relationships, and your ability to reason with numbers?

- How might you use this routine in your classroom?

Inspired by the Tell Me All You Can routine in *Math Workshop Essentials* (Bresser & Holtzman, 2018, pp. 3–9)

Manipulatives and Models Matter: Manipulative Models and Operation Sense

When we were in elementary school, many of us were taught to use the "key word" strategy to solve word problems. Lists of key words for each operation were often posted in the classroom (Addition: altogether, total; Subtraction: left, take away; Multiplication: each, equal groups; Division: split, share). We were instructed to look for these key words in a word problem as if on a scavenger hunt. These key words, we were told, would let us know which operation we should use to solve the problem.

We know now that the key word strategy does not help students to develop mathematical competence and that it damages students' math identities and agency (SanGiovanni et al., 2022; Van de Walle et al., 2019). The key word strategy is ineffective because:

- **Key words can be misleading.** Consider this problem: There are three boxes of markers on the table. There are eight markers in a box. How many markers are there altogether?

- **Some problems don't include key words.** Consider this problem: Tajra wants to buy a movie ticket for $5.00. She has $3.25. How much money does she need?

- **When students rely on the key word strategy rather than making sense of the problem itself, they are not equipped to tackle multiple-step problems.** Consider this problem: Roch read for 13 minutes on Monday, 11 minutes on Tuesday, and 18 minutes on Wednesday. How far is he from reaching his goal of reading a total of 60 minutes this week?

The most important reason, however, for choosing *not* to teach students to rely on key words is that this strategy destroys students' confidence in themselves as mathematical sense-makers. A focus on key words communicates to students that mathematics is about tricks rather than about reasoning and sense-making. We need to help students use all the words in a word problem, make sense of the meaning of the problem, and then figure out the mathematical actions that can be used to solve it.

Manipulative models are a powerful tool for helping students to make sense of word problems and for building their operation sense (Van de Walle et al., 2019). Instead of teaching students to rely on key words, we can instead have students use manipulative and pictorial models as thinking tools to make sense of word problems and suggest solution strategies.

Try this experience with your math teaching community:

1. Use the Fist Bump Problem or another word problem from your math curriculum.

2. Individually choose a math manipulative to model this problem.

3. Compare and discuss the various ways team members chose to model the problem.

REFLECT

*   How did the use of a manipulative model help you to make sense of the problem?

*   How did it help you to think about the operation or operations you might use to solve this problem? How did this experience deepen your understanding of these operations and their potential uses?

*   How can you create opportunities for your students to make sense of word problems and build their operation sense through the use of manipulative models?

Game Time

This month's game offers a great way to build number sense, as well as meaningful practice of computational skills. It can be easily adapted to a variety of math content focuses and for differentiation purposes. It can be played whole class (teacher vs. the class), in partners, or in groups of three or four.

THE GREATEST WINS

Materials:

2–4 players

One die

Game board drawn on paper or whiteboards

Pencil or dry-erase marker

(Continued)

(Continued)

Challenge: Be the player with the greatest final answer.

Game directions:

1. Choose a game board appropriate to your grade level's current curriculum focus. Each player draws a copy of the chosen game board on a piece of paper.

2. On your turn, roll a die. Write the number rolled in the box of your choice. The reject box may be used once for a number you don't wish to use for game play. Once you write a number down, you may not change it.

3. Pass the die to the next player. This player rolls the die and records the number rolled in the box of their choice.

4. Continue taking turns rolling the die and recording numbers until all boxes have been filled. Perform the calculation if needed.

5. The player with the greatest final answer wins the round.

Bonus step to support operation sense: The winner verbally shares a word problem that goes with the calculation performed.

Discuss the experience:

- How did you draw on your number sense as you decided where to write each number rolled on the die?

- How can the use of games like *The Greatest Wins* build students' number sense and their mathematical agency?

- What math habits did you use while playing *The Greatest Wins*?

- How might you use or adapt *The Greatest Wins* for use with your students?

Inspired by The Greatest Wins activity in *About Teaching Mathematics* (Burns, 2022, pp. 94–96)

 Available as a downloadable resource on the companion website.

IN YOUR CLASSROOM

Classroom Story

Ms. Sergeant asked her students to retrieve their table baskets of math tools for a warm-up problem. She pointed to the word problem on the board:

> The Villa family had a picnic in the park. There were 3 picnic tables, 4 people were sitting at each table. How many people were at the picnic?

The class was just beginning their study of multiplication. Ms. Sergeant wanted students to experience and make personal sense of contexts involving multiplication before introducing the vocabulary and symbolism connected with this operation. She planned to have students model the problem with their choice of manipulatives so they could see the equal groups and recognize the need to join these groups to find a total.

Ms. Sergeant read the math story aloud. She asked students to talk briefly about their experiences with family picnics and picnic tables to make sure they understood the context of the problem. Then she asked students to retell the story to a shoulder partner. After a minute, she invited a few students to share their retellings with the class.

Ms. Sergeant asked a couple of questions to ensure that students understood the problem. "What does 'at each table' mean?" "What are we figuring out in this problem?"

She then asked students to begin working on the problem.

Liddy used three long rods from the base ten materials to represent the three picnic tables. She carefully placed four unit blocks at each table, two on each side.

Xander arranged four color tiles in a square and then repeated this arrangement two more times. Other students used two-color counters and connecting cubes to think about and solve the problem.

As Ms. Sergeant walked around, she checked in with students who seemed hesitant to start. She wanted to nudge students to take initiative in making sense of the problem and to understand how students were thinking about this mathematical situation. Ms. Sergeant had preplanned a couple of questions with these purposes in mind. She'd jotted these questions on an index card that she referred to as she interacted with students at work.

- *What is this problem about? How can you show that with a math tool?*

- *How do your manipulatives represent this story? How do they help you to figure out the answer to the problem?*

After a few minutes, Ms. Sergeant brought the class back together. She invited several students to share their representations on the document camera and to talk about how the models helped them to make sense of the problem.

Ms. Sergeant asked students to clean up their manipulatives and join her on the carpet for math workshop time. As the students put their materials away, Ms. Sergeant reflected on how student talk and choice in how to represent the problem had positioned her students as mathematical sense-makers.

Anchor Lesson

Self-Questioning to Support Sense-Making

WHAT?

This lesson will build students' awareness of when a mathematical idea does or doesn't make sense and allow students to practice using self-questioning to support sense-making.

YOU NEED:

Challenging math task

Chart paper

Markers

HOW?

1. Write the words "Making sense" on an anchor chart. Verbally give a couple of examples of times when you made sense of a mathematical idea or problem. Invite students to talk about what "making sense" means and to give examples of times when they made sense of something.

2. Tell students that one way we can help ourselves to make sense of things is to ask ourselves questions. Write these questions on the anchor chart:

   Does this make sense? Why or why not?

   Tell students that they will engage in a problem-solving experience and that every once in a while you will pause the class so that students can ask themselves these questions. Encourage students to notice how reflecting on the questions impacts their math thinking and agency.

3. Engage students with a **thinking-rich mathematics problem**. You might choose a task from your mathematics curriculum or try out one of these problems.

   How many ways can you solve this problem?

   Option 1: __ + __ + __ = 10

   Option 2: __ – __ = 17

   Option 3: __ × __ = 1,875

   Option 4: __ ÷ __ - 5

   As students work, periodically stop the class and give students a minute to reflect on the questions "Does this make sense? Why or why not?" After students have had a minute to think, ask them to share their thinking with a turn-and-talk partner.

4. During the class discussion at the end of the lesson, ask students to talk about how they felt when they asked themselves these sense-making questions and how the questions helped them to think about and make sense of mathematical ideas.

5. As a class, add some action steps to the anchor chart you started at the beginning of class. You may even have students provide some ideas for making sense and include them on the anchor chart. Here are some ideas to get you started:

   • Stop frequently to check in and ask the questions, "Does this make sense? Why or why not?"

(Continued)

(Continued)

- Pay attention to confused feelings. If you have them, try these action steps:
  - Reread the problem
  - Talk with someone else

REFLECT

When students are active thinkers, they seek to make sense of math rather than passively following a series of steps. We can teach students to be metacognitive about their thinking, ideas, and strategies while they problem-solve.

By asking students to stop and ask questions, we can prepare them to practice the habit of sense-making.

Use the discussion questions to reflect with students:

- Why is it important to make sense of math?
- How will stopping to ask yourself the question "Does this make sense?" help you in doing and learning math?

LET'S DO SOME MATH ACROSS OUR SCHOOL!

The Fist Bump Problem is a great way to build your classroom math community. It allows students to interact in a fun way, think about the mathematics embedded in simple daily activities such as greeting classmates, and practice mathematical sense-making and other math habits. It is accessible for kindergarten students but has plenty of complexity for students at all grade levels, as well as for adult mathematicians.

Pose the problem: If everyone in our class gave everyone else a fist bump, how many fist bumps would there be?

You might start with a simpler problem, perhaps thinking about the number of fist bumps in a group of four students. You'll likely want to act the problem out. Explore different ways of keeping track of and representing the number of fist bumps for different group sizes: manipulatives, diagrams, tables, and equations. Be sure to capture your students' math thinking on a class chart or have students journal about their math thinking after you discuss the problem as a class. Post these records of your students' mathematical brilliance outside the classroom where others can admire and be inspired by your student mathematicians.

The Fist Bump Problem enchants mathematicians of all ages. It invites us to notice and talk about patterns, speculate about why they occur, and generalize how these patterns would play out with larger and larger groups of people. You might work through the basic problem with your class one day but then leave a teaser: "How many fist bumps will there be when your family is all together?" or "I wonder how many fist bumps there would be if our class and Ms. Hinojosa's class met for recess." Follow your students' passion. The Fist Bump Problem is a great problem to think about with your class across a week, a month, or longer.

Perhaps the fist bump investigation will lead your class to think about other ways that mathematics is embedded in the act of greeting others. You might decide to tally or graph the different ways that students like to be greeted at the classroom door each morning by the teacher (high five, fist bump, handshake, etc.). Or you could investigate the number of times class members greet or acknowledge others by name to build students' communication skills and a sense of community.

TEACHING MOVE: HAVE STUDENTS WORK SILENTLY AND INDIVIDUALLY BEFORE COLLABORATING

We know that students benefit from collaborating on challenging mathematics problems. The opportunities to articulate math thinking for others and to hear classmates' ideas and strategies can accelerate and deepen learning (Chapin et al., 2022). But when students work together, the more extroverted and confident students sometimes take over, not leaving room for the voices or ideas of quieter and less confident students. As a result, quieter students lose the opportunity to practice sharing their math thinking and the group misses the chance to learn from these students' contributions and unique ways of thinking. Providing individual think time before collaboration helps all students to make personal sense of a problem so they can enter collaborative spaces ready to contribute to the group's thinking and problem-solving work.

When preparing students to engage with a thinking-rich mathematics problem, follow this sequence of steps:

- **Whole class**: Activate the prior knowledge students need to make sense of the problem. Ensure that students understand the problem and know what they are trying to figure out.

- **Independently**: Ask all students to work silently and individually on the problem for a few minutes to make personal sense of the problem and begin generating ideas for solution strategies.

- **Partners or small groups**: Once you see students moving from sense-making into solving, invite students to work together in partnerships or small collaborative groups to continue investigating and solving the problem.

- **Whole class**: Facilitate a class discussion of strategies used and the mathematical ideas embedded in the problem.

All students need and deserve the opportunity to make personal sense of a problem before collaboration begins. Sense-making takes time. We must, therefore, build this thinking time into the design and implementation of our mathematics learning activities.

REFLECT

- How does giving students individual think time before collaborating impact the number of students who actively collaborate in group work and the groups' depth of thinking?

TEACHING MOVE: TEACH STUDENTS TO USE ESTIMATION AS A SENSE-MAKING STRATEGY

Estimation is an underused and often misunderstood mathematical skill. When asked to complete an estimation problem on an assessment, we sometimes see students first calculate the exact answer to the computation and then round that answer to come up with what they think is an estimate. They are missing the purpose of estimation, to calculate a good enough answer as easily and efficiently as possible.

Estimation is a valuable tool for doing real-life mathematics. According to Burns (2022), estimates, rather than exact answers, are sufficient and often more appropriate for about half of the arithmetic we do on a daily basis (p. 27). Think about it—we estimate tips, discounts, mileage, and time. Frequently, we need only ballpark answers to the mathematical questions we ask ourselves as we navigate everyday life.

Estimation is an invaluable strategy for making sense of mathematical situations. We can encourage students to estimate:

- Before computing the exact answer to a numerical problem.

- After computing the exact answer to a numerical problem.

Estimating before computing supports the development of number sense. Estimating after computing develops the ability to assess the reasonableness of answers. As students develop the habit of automatically using estimation as a sense-making strategy, their mathematical confidence and agency grow stronger.

REFLECT

- How does the use of estimating as a sense-making strategy build your students' confidence in their mathematical abilities?

TEACHING MOVE: POSITION STUDENTS AS SENSE-MAKERS BY SHARING LEARNING POWER

For students to see themselves as capable of making sense of mathematics, we need to move past the traditional view of the teacher as the expert who imparts knowledge and of students as passive recipients of this knowledge. In a classroom where students see themselves as sense-makers, the power to construct knowledge is shared between the teacher and students.

Here are some ways to share learning power with students to make room for sense-making:

- Ask open-ended questions that have multiple correct answers.

- Use tentative language such as *how might*, *what if*, *I wonder*.

- Provide a generous wait time to allow for thinking.

- After a turn-and-talk partner discussion, ask students to share their partner's thinking rather than their own.

- During class discussions, encourage students to respond to classmates' ideas, addressing each other directly rather than responding to you.

- When possible, hold class discussions in a circle on the floor so that all class members including you are physically at the same level and facing each other.

- Teach students how to facilitate a class discussion and provide opportunities for students to take on this role. (Johnston, 2012, p. 56)

We empower our students intellectually and mathematically when our teaching actions reflect our belief that all students are mathematical sense-makers.

REFLECT

- How do teaching moves that share learning power set students up for future success in mathematics?

CLASSROOM ROUTINE: NOTICING AND WONDERING

The noticing and wondering routine is a simple but powerful classroom routine to build students' sense-making abilities. Max Ray, author of *Powerful Problem Solving* said that noticing and wondering helps students to:

- Connect their own thinking to the math they are about to do

- Attend to details within math problems

- Feel safe (there are no right answers or more important things to notice)

- Slow down and think about the problem before starting to calculate

- Record information that may be useful later

- Generate engaging math questions that they are interested in solving

- Identify what is confusing or unclear in the problem

- Conjecture about possible paths for solving problems

- Find as much math as they can in a scenario, not just the path to the answer (2013, p. 46)

Use the following steps to get your class started with noticing and wondering:

1. Share a word problem without the question at the end. Omitting the question helps students to think deeply about the problem scenario. It also encourages students to pose their own mathematical questions.

2. Ask, "What do you notice?" Allow wait time for thinking and then call on as many students as possible to share their ideas. Record these noticings on the board.

3. Ask, "What do you wonder?" Again, allow wait time and then call on multiple students to share, recording their wonderings as questions on the board.

4. Reveal the question that was part of the original problem and have students solve the problem.

(Continued)

(Continued)

Variations:

- Instead of a word problem, show an image that involves mathematical ideas. Elicit noticings and wonderings. Allow students to investigate a wondering that interests them.

- Use noticing and wondering as a think–pair–share prompt to promote engagement and accountability.

- Encourage students to use noticing and wondering as a personal strategy for getting unstuck during problem-solving.

Ray said that the noticing and wondering routine produces dramatic results in students' sense-making and learning. He explained:

> Learning depends on mulling, connecting, wondering, and repeatedly thinking about, and noticing and wondering enables this to take hold and blossom. When we invite everyone to share in math class, and we see how each student's contribution builds toward a complete mathematical understanding of the problem at hand, we invite students to think of themselves as mathematicians. (2013, p. 55)

STATION ACTIVITY:
MY SENSE-MAKING STORY

This station activity is designed to help students become more self-aware of when things don't make sense and build their agency for making sense of mathematics.

Think about a time when you were confused about a mathematical idea or problem (in the pit!), but eventually you made sense of this mathematics. Write your sense-making story or create a storyboard, drawing pictures to show what happened.

What was confusing?	How did I think about it?	What did I learn?

Post your story or storyboard for others to read.

Read your classmates' stories and leave them encouraging feedback on a sticky note using this sentence stem:

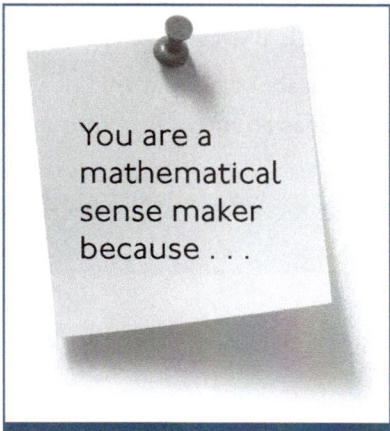

You are a mathematical sense maker because . . .

Source: Istock.com/Thammask-Chuenchom

 Available as a downloadable resource on the companion website.

LITERATURE CONNECTIONS

***Lia & Luís: Who Has More?* by Crespo** (2020) (Grades K–2)

Lia and Luís each receive a bag of their favorite Brazilian snack. Lia chooses coxinhas de galinha, and Luís selects biscoito de polvilho. The twins wonder who received more. They investigate their question by applying mathematical ideas of quantity and measurement.

Activity: Lia and Luís use mathematics to make sense of something they are wondering about. What is something you are interested in? How can you use math to make sense of this thing and to learn more about it?

***If the World Were a Village: A Book about the World's People* by Smith** (2020) (Grades 3–5)

David Smith's mission to help children become "world-minded" led him to create this thought-provoking book celebrating the diversity of people that exists in our world. If there were just 100 people instead of 700 million people in the world, 59 would live in Asia, 16 would live in Africa, 10 would live in Europe, 9 in South America, 5 in North America, and 1 in Oceania. Smith shared additional statistical comparisons for languages, ages, religions, food, air and water, school and work, money and possessions, energy, health, and history. He offered a variety of activities for extending this learning to help students gain an understanding of our world and their place in it.

Activity: In this book, the author used math to help us make sense of our world and its people. What is one idea from this book that surprised you? How can you gather and represent some mathematical facts about people who live near you to help you and others make sense of important ideas about your community?

SPOTLIGHT ON BRAIN SCIENCE

When we are curious, a part of our brain called the "reticular activating system" looks for input to satisfy that curiosity. This focused search for input results in learning (Willis & Willis, 2020). We want students to be curious and to know that their curiosity leads to learning. Students can learn to activate their curiosity through noticing and wondering and by asking themselves sense-making questions.

You can share this information with students in a mini lesson or a station activity. You might also send this reproducible home along with the Family Newsletter to help families talk about these important ideas.

CURIOSITY HELPS US LEARN.

Babies are born naturally curious about their world. This is why they learn so much. When we are curious about something, our brains look for information to satisfy our curiosity.

You can boost your math learning by choosing to be curious about mathematical ideas. This starts with noticing and wondering. Before you solve a math problem, think, "What do I notice? What do I wonder?" Notice things that are interesting about the problem and ask yourself questions that help you to think about the problem more.

A good curiosity question to ask yourself about mathematical ideas is "does this make sense?" If the idea does make sense, explain why it makes sense to someone else or even to yourself. If it doesn't make sense, think about what is puzzling and why the idea doesn't make sense. When you ask "does this make sense?" you activate your mathematical curiosity and learning.

You can practice noticing and wondering with anything that you are interested in. What do you notice and wonder about this magnetic resonance image (MRI) of a human brain and eye?

(Continued)

(Continued)

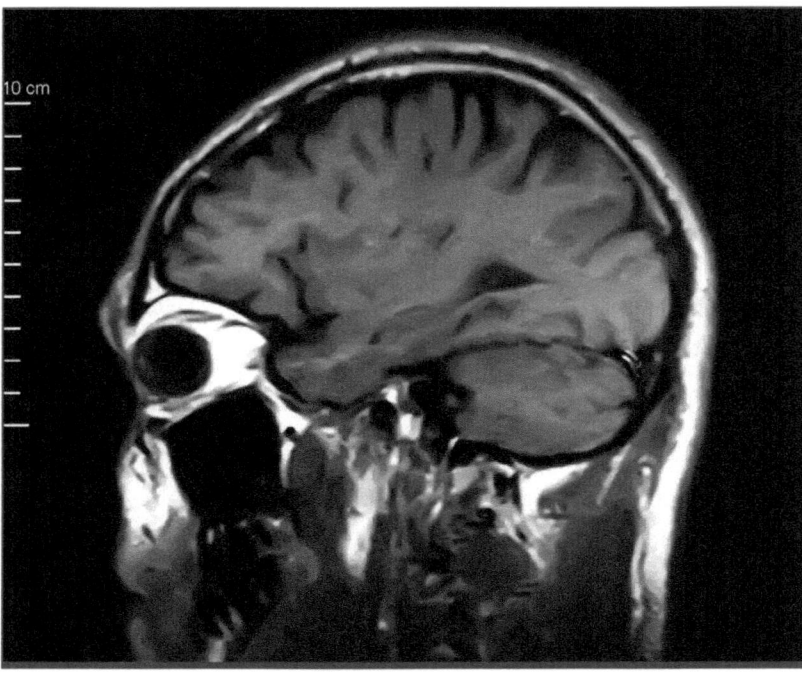

Image source: Istock.com/Juan-Ruiz-Paramo

Choose one of these ways to think more about this important information:

1. Write down why these ideas are important to you and how you want to put them to work in your life.

2. Talk to a friend about these ideas. Decide together how you will put them to work.

3. Share these ideas with someone at home. Tell them why these ideas are important to you and how you plan to use them.

 Available as a downloadable resource on the companion website.

SPOTLIGHT ON EQUITY: SUPPORT ALL STUDENTS IN BECOMING INDEPENDENT LEARNERS

Zaretta Hammond, author of *Culturally Responsive Teaching & the Brain* (2014) said that one way we can promote equity in our classrooms is by helping students who are currently **dependent learners** learn how to be **independent learners**. Dependent learners don't see themselves as capable of making sense of mathematics. They don't expect math to make sense. Independent learners take initiative in their learning because they expect math to make sense and they see themselves as capable of mathematical sense-making. Consider Hammond's descriptors of both types of learners (2014, p. 14) in Table 3.2 and think about how you have seen these behaviors in students, past and present.

Table 3.2 • Characteristics of Dependent and Independent Learners

The Dependent Learner	The Independent Learner
• Is dependent on the teacher to carry most of the cognitive load of a task always	• Relies on the teacher to carry some of the cognitive load temporarily
• Is unsure of how to tackle a new task	• Utilizes strategies and processes for tackling a new task
• Cannot complete a task without scaffolds	• Regularly attempts new tasks without scaffolds
• Will sit passively and wait if stuck until teacher intervenes	• Has cognitive strategies for getting unstuck
• Doesn't retain information well or "doesn't get it"	• Has learned how to retrieve information from long-term memory

Source: Reprinted with permission from Hammond, 2014, p. 14.

Independent learners have agency. They recognize the necessity of being "in the pit" as they make sense of complex mathematics, but they have confidence in their ability to climb out of this cognitive dissonance. Hammond stated, "We have to help dependent students develop new cognitive skills and habits of mind that will actually increase their brainpower" (2014, p. 15). The math habits are math-specific cognitive skills and habits of mind. When we help students to strengthen their math habits, we enable them to grow mindsets and skills needed for independent learning.

TRY IT

Think of a student who exhibits some of the behaviors of dependent learners. Set aside a brief time when you can conference with this student about what it means to be an independent learner and why self-directed learning behaviors are important. Support the student in setting a small goal toward growing the skills of independent learning. Together create a plan for specific things the student and you will each do work toward achieving this goal. Schedule a time to meet again to monitor progress.

MATHEMATICAL ME: STUDENT JOURNAL AND PORTFOLIO

Student Journals:

Have students respond to the following questions:

- Do I expect math to make sense?

- What should I do when a math idea doesn't make sense to me?

Older students can write their responses in their Mathematical Me journals. Younger students can draw a picture showing their response, or you might gather responses during a class discussion or quick one-on-one interviews.

Review students' responses to monitor students' beliefs about making sense of mathematical ideas. You might summarize this data by tallying or graphing the different ideas students offer.

Be sure to share and discuss the class-level data minus student names with your class. Ask, "What does this data tell us?" and "What are some things we can do to continue growing as mathematicians?"

Student Portfolios:

Have students choose a piece of work where they made sense of a math problem. On a sticky note, have them write the date and complete this sentence frame:

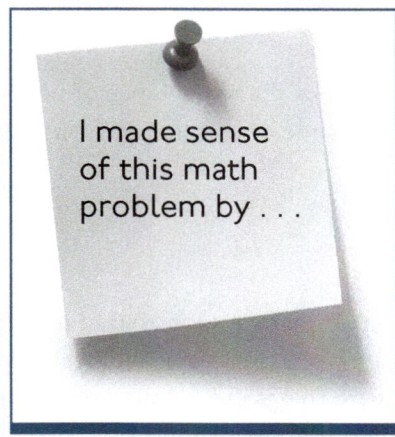

I made sense of this math problem by . . .

Source: Istock.com/Thammask-Chuenchom

FAMILY NEWSLETTER: WE EXPECT MATH TO MAKE SENSE!

WHAT WE'RE LEARNING

This month we're learning about the importance of believing that math is supposed to make sense and that every one of us can make sense of math and learn math with understanding.

WHY IT'S IMPORTANT

When your child believes that math is supposed to make sense, they take an active role in thinking about and figuring out mathematical ideas that may be confusing at first. This belief that we can figure things out is called *agency* and it helps your child to be a confident, self-directed learner.

HOW YOU CAN HELP

Here are two ideas for helping your child develop their math sense-making abilities and agency.

1. Encourage mathematical curiosity. In school, we practice a routine called *noticing and wondering*. We take a minute to notice interesting features about a math problem and then another minute to pose questions or wonder mathematically about this same problem. You can practice noticing and wondering at home with things that you and your child observe together. For instance, if you observe a flock of birds flying overhead as you take a walk, you might generate statements like these together:

    *   I notice that the birds are flying in the same direction.
    *   I notice that the flock of birds is shaped like a triangle.
    *   I notice that each bird is longer from wingtip to wingtip than from head to tail.
    *   I wonder how many birds there are in this flock.
    *   I wonder where these birds came from and how far they will fly.
    *   I wonder how often they stop to rest and eat.

2. Practice computational estimation. Estimation is a powerful sense-making skill because it activates critical thinking and number sense. You can strengthen your

child's estimation skills and confidence by estimating amounts and measurements together related to daily activities. Here are some examples:

- If a sandwich costs $2.59 and a drink costs $1.59, about how much will lunch cost?

- How much milk do we as a family drink each day and how much should we buy at the store every week?

- Grandma's house is about 350 miles from our house. If we go for a visit by car, what is our average speed likely to be and about how long will it take to drive there? How often will we need to stop for breaks?

 Available as a downloadable resource on the companion website.

CHECKING IN ON OUR LEARNING: WE EXPECT MATH TO MAKE SENSE!

- Do I expect math to make sense?
- What should I do when a math idea doesn't make sense to me?

This month's learning focus, expecting math to make sense, is foundational to the other math habits. When students expect math to make sense and believe that they can make sense of math, they are ready to take on challenging mathematics. Engaging in challenging math experiences will, in turn, grow students' mathematical understandings, skills, and confidence, setting in motion an iterative cycle of self-directed math learning. In a powerful math community, educators value and support the development of their students' mathematical agency alongside their mathematical competence.

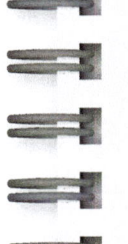

MATHEMATICAL ME: EDUCATOR JOURNAL

- How has this month's learning focus supported your students' mathematical growth?

- How has this month's learning focus supported your growth as a math teacher?

- How has this month's learning focus supported your school's growth as a powerful math community?

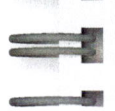

LOOKING AHEAD

In October, we will look at the role of struggle and perseverance in learning. We'll consider strategies mathematicians can draw on to push through struggle and how educators can design learning experiences to help students develop personal toolkits of perseverance strategies.

Questions for Reflection

- When you encountered challenging math problems in school, how did you respond? Why?

- What are some ways your students respond to challenging mathematics problems? How would you like them to respond?

- What does our response to challenging math problems have to do with math learning?

CHAPTER FOUR

October
We Love Challenging Math Problems!

ESSENTIAL QUESTIONS

- What is productive struggle, and why is it important?
- What are strategies I can use to persevere with challenging problems?

THIS MONTH'S FOCUS

In a powerful math community, students and educators know that struggle is essential to learning. But they also know that too much struggle can impede learning. They have, therefore, developed a personal toolkit of strategies for turning unproductive struggle into productive struggle.

Educators intentionally design learning experiences to provide students with opportunities for productive struggle. They strategically use learning supports to help students grow an array of strategies for persevering with challenging math problems.

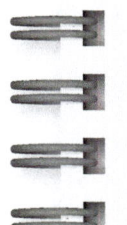

MATHEMATICAL ME: EDUCATOR JOURNAL

Take five minutes to respond to the essential questions. You will have a chance to reflect on these same questions and to have students respond to these questions at the end of the month.

ON YOUR OWN

LET'S DO SOME MATH! THE HORSE PROBLEM

Math education guru Burns (2015b) famously uses the Horse Problem to help educators notice our cognitive process as we engage with complex math problems and how our emotions come into play during the problem-solving process. The Horse Problem is simple to understand but also challenging to think about.

A person bought a horse for $50 and sold it for $60.

This same person bought the horse back for $70 and sold it again for $80.

Did the person make money or lose money? How much did they make or lose?

Image source: Istock.com/cattallina

Give it a go! And as you tackle this problem, notice how you engage in the habits of mathematically powerful people. Also, pay attention to your emotions and how they change throughout the problem-solving process.

online resources ⬄ Available as a downloadable resource on the companion website.

Why This Focus

How did it feel to wrestle with this problem? Did you struggle? Was it a fun kind of struggle or a frustrating kind of struggle? What does struggle have to do with learning?

Sue's Math Story:

The preservice teaching candidates in my university math methods class come into our course with a variety of experiences and feelings about mathematics and math learning. Their first assignment of the semester is to create a video in which they reflect on their experiences learning math in elementary school and consider how these experiences may impact their future math teaching. As I listen to their videos, I frequently hear candidates say things like, "I always struggled with math" or "I don't want my future students to struggle the way I did."

Our beliefs and assumptions about "struggle" are worth examining because they influence the learning experiences we design for students and our in-the-moment support for learners. Gonzalez told us that struggle is essential to math learning. She said, "Students who are most successful [in math] are those who can sit with the discomfort of not knowing and who are able to continue to work through problems regardless. Yet struggle is not what comes to mind when most people consider what it means to be good at mathematics" (Gonzalez, 2023, p. 14). If learning is understood as the process of making personal sense of new ideas and then assimilating these new ideas into our existing schema, then struggle is a natural and needed part of learning work.

Struggle, however, is only desirable if learners push through it rather than giving up. We call this **productive struggle**. SanGiovanni et al. (2020), authors of *Productive Math Struggle*, defined productive struggle as "purposefully reacting to an unclear challenge so that progress is made or learning advanced" (p. 28). In contrast, when struggle causes learners to give up, when they don't have the internal resources or external support to push through struggle, struggle becomes unproductive. **Unproductive struggle** does not lead to learning, and it can nibble away at a learner's self-efficacy and agency.

> If learning is understood as the process of making personal sense of new ideas and then assimilating these new ideas into our existing schema, then struggle is a natural and needed part of learning work.

Tapping Into Your Experience

Think of a specific time when you experienced struggle as a math learner. Zoom into that moment as closely as possible.

- What do you remember about the mathematics you struggled with? What emotions did you experience? Who else was present, and what role did they play in this experience?

- Were you able to persevere through the struggle, or did you stop trying? If you gave up, what internal or external supports might have helped you to persevere?

- How did this struggle experience impact your long-term feelings about mathematics and your identity as a math learner? Consider how this struggle experience may influence how you think about math learning today and how you currently support your students as math learners.

Because productive struggle is essential to learning and unproductive struggle can be destructive, teachers need a toolkit of specialized struggle-management skills, including:

- The ability to recognize the signs of productive and unproductive struggle

- The ability to design learning experiences that promote and support productive struggle

- A repertoire of scaffolding strategies to help students turn unproductive struggle into productive struggle

These skills are not commonly taught in teacher preparation programs. They can be introduced in professional learning settings but are developed and refined in the classroom. We continue to strengthen these essential skills across our teaching careers. However, this important professional learning can be accelerated when school-based teams work collaboratively to grow this skill set and when teachers are provided with feedback about their use of these skills from school leaders, instructional coaches, and peers. You will find activities to help you and your colleagues develop your toolkits of struggle-management practices throughout this chapter.

> Because productive struggle is essential to learning and unproductive struggle can be destructive, teachers need a toolkit of specialized struggle-management skills.

Is It Productive or Unproductive Struggle?

If we, as educators, are to support productive struggle and help students move out of unproductive struggle into productive struggle, we need to be adept at recognizing signs of both productive and unproductive struggle. SanGiovanni et al. (2020, p. 19) offered this list of behaviors that are commonly observed during unproductive and productive struggle (see Table 4.1). As you read through these examples, think about times you have noticed these behaviors in students and what they might indicate about a student's agency as a math learner at that moment. What other behaviors might you add to this list?

Table 4.1 • Student Behaviors Associated With Mathematical Struggle

	BEHAVIORS THAT MAY SIGNAL STRUGGLE AVOIDANCE	BEHAVIORS THAT MAY SIGNAL STRUGGLE DEFLECTION	BEHAVIORS THAT COULD SIGNAL STRUGGLE FRUSTRATION
Unproductive Struggle	• Asking off-task or unrelated questions • Going to the bathroom, getting a drink, sharpening a pencil, going to the nurse • Playing inappropriately with materials • Putting head down • Copying others' work • Talking with classmates	• Making jokes, silliness, teasing • Asking unrelated questions, sharing unrelated stories • Being rough with materials • Acting out	• Arguing with statements or directions • Acting out • Breaking pencils, being rough with materials • Teasing classmates • Looking at what others are doing • Putting head down • Tearing up, crying
	BEHAVIORS THAT MAY SIGNAL STRUGGLE TO UNDERSTAND THE PROBLEM	BEHAVIORS THAT MAY SIGNAL CONSIDERATION OF POSSIBLE STRATEGIES TO USE	BEHAVIORS THAT MAY SIGNAL HOW STUDENTS ASSESS THEIR PROCESS OR STRATEGY
Productive Struggle	• Asking on-task or related questions • Using manipulatives • Drawing a representation of the problem	• Getting up to get manipulatives or tools • Talking with a classmate • Asking a classmate for assistance • Walking around the room	• Walking around the room • Looking at reference materials (journals, charts) • Using calculators or other tools • Talking with classmates

Source: Reprinted with permission from SanGiovanni et al., 2020, p. 19.

 Available as a downloadable resource on the companion website.

Challenge, Struggle, Safety, and Growth Mindset

If struggle is so important to learning, what do we do when a student avoids challenging mathematics that leads to struggle?

Boaler reminded us that a safe classroom environment is prerequisite to academic risk taking. She wrote:

> Not only should the work be challenging to foster mistakes; the environment must also be encouraging, so that the students do not experience challenge or struggle as a deterrent. Both components need to work together to create an ideal learning experience. (2019, p. 49)

Across the year, teachers nurture a classroom environment in which productive struggle and growth can occur. Self-fulfilling prophecy, a strengths-based perspective and language (see Chapter 2—August), the expectation that math is supposed to make sense and that I am capable of making sense of math (see Chapter 3—September), and the recognition of mistakes as a natural and needed part of learning (see Chapter 5—November) are all integral to a classroom in which productive struggle and high levels of math learning take place on a regular basis.

Boaler (2019) also told us that "when students don't want to struggle, it is because they have a fixed mindset; at some point in their lives they have been given the idea that they cannot be successful and that struggle is an indication that they are not doing well" (p. 61). The concepts of challenge, struggle, safety, and mindset are all linked. Students with a **growth mindset** value struggle because they know it is essential to their growth. Such students view intelligence, skills, and abilities as improvable over time and with effort—and thus value effort—whereas students with a **fixed mindset** view these same traits as unchanging over time. Students with a growth mindset frame struggle as challenge, an opportunity to stretch beyond their current capabilities. Similarly, when students practice opening their minds and their hearts to challenge and struggle, they strengthen the habit of approaching learning with a growth mindset.

> Engaging students in challenging mathematics, teaching them about productive struggle, and helping them learn to value struggle are important vehicles for developing our students' growth mindsets.

Our students' mindsets toward learning are critical to their success in school and life. In his TEDx Talk *The Power of Belief—Mindset and Success,* growth mindset expert Eduardo Briceno said:

> How is it possible that, as a society, we are not asking schools to develop a growth mindset in children? Our myopic effort to teach them facts, concepts, and even critical thinking skills is likely to fail if we don't also deliberately teach them the essential beliefs that will allow them to succeed not only in school but also beyond. (TEDxManhattanBeach, 2012; https://bit.ly/3V1P9c5)

As educators, we must help our students adopt a growth mindset toward learning. Engaging students in challenging mathematics, teaching them about productive struggle, and helping them learn to value struggle are important vehicles for developing our students' growth mindsets.

Teachers need to know how to teach students to persevere when learning is challenging. **Perseverance** allows the learner to push through struggle into understanding; it can be thought of as self-scaffolding. Perseverance is an important ability for all students to have in learning mathematics, and it is extremely important for navigating life.

It is a skill set, and with practice, perseverance can become a habit. We can support students in learning to persevere through these teacher actions:

- Design learning experiences that allow for struggle. Select and use complex thinking-rich mathematics tasks and problems.

- Provide opportunities for students to reflect individually and as a class on the experience of struggle in mathematics and to name emotions experienced as a result of that struggle.

- Explicitly teach and then practice strategies for persevering through struggle. Provide actionable feedback and coaching to students as they practice perseverance strategies.

- Teach students how to monitor the need for perseverance strategies and self-assess their growing ability to persevere.

See the *Anchor Lesson* in the *In Your Classroom* section of this chapter (p. 121) for an example of how teachers can introduce their students to the concept of perseverance.

Mathematics learning is maximized when both teachers and students come to value and enjoy productive struggle. We can cultivate this "bring it on!" attitude toward challenging mathematics when we accept responsibility for teaching our students the importance of struggle and perseverance.

Characteristics of Math Problems That Promote Struggle

Liljedahl (2021), author of *Building Thinking Classrooms in Mathematics, Grades K–12*, stated that "good problem-solving tasks require students to get stuck and then to think, to experiment, and to try and to fail, and to apply their knowledge in novel ways in order to get unstuck" (p. 20). In other words, good problem-solving tasks are challenging and give students opportunities to productively struggle.

Kobett and Karp offered the following characteristics of thinking-rich math problems:

- I don't know how to solve the problem right away.
- I need patience to solve the problem.
- I need to try a few different strategies.
- I may need to start the problem over or work backwards.
- I may need to talk about my idea with a partner. (2020, p. 49)

Find an example of a mathematics problem that has these characteristics. Try it out with your class. As students work on the task, watch and listen for productive struggle using Table 4.1.

Boaler's website Struggly offers a selection of thinking-rich mathematics tasks designed to engage students in productive struggle: https://joboaler .org/struggly/.

Strategies for Modifying Math Problems to Promote Struggle

Many mathematics tasks found in curriculum resources offer students little challenge and, therefore, no opportunity for productive struggle. Boaler (2022, p. 92) suggested six strategies for bumping up the thinking level and learning value of the math problems we use with students. We have added the following examples.

1. **Open the task to encourage multiple methods, pathways, and representations.**

   Original problem: 8 + 27

   Modified problem: How many different strategies can you come up with to mentally add 8 to 27?

2. **Include inquiry opportunities.**

   Original problem: Which of these numbers are odd? (327, 48, 99, 501, 760)

   Modified problem: Five numbers are added together. Their sum is an odd number. What do you know about the five numbers?

3. **Ask the problem before teaching the method.**

   Example: (before teaching a procedure for multiplication of two-digit by one-digit numbers) We'll ride the bus to the art museum for our field trip next week. We need to make sure there is enough room for everyone on the bus. There are 12 rows of seats on the bus, and six students can sit on each row. Represent this problem in a way that makes sense to you and figure out how many students can ride on the bus.

4. **Add a visual component and ask students how they see the mathematics.**

   Original problem: How many ways can you decompose 25? Show your thinking with equations.

   Modified problem: Use base-ten blocks to show all the ways to make 25. Record each way with an equation. Which way uses the fewest blocks? Which way uses the most pieces? Why?

5. **Extend the task to make it lower floor and higher ceiling.**

   Original problem: I have two quarters, two dimes, four nickels, and 10 pennies in my pocket. How much money do I have?

   Modified problem: I have $1.00 in coins in my pocket. What coins might I have?

6. **Ask students to convince and reason; be skeptical.**

   Original problem: If four people share six cookies equally, what is each person's share?

   Modified problem: Siara, Elise, and Solomon solved this problem. If four people share six cookies equally, what is each person's share?

   Siara said the answer to the problem was $1\frac{2}{4}$ cookies.

   Elise said the answer to the problem was 1.5 cookies.

   Solomon said the answer to the problem was $1\frac{1}{2}$ cookies.

   Who is correct? How do you know?

Take a low-level mathematics problem and make it more rigorous by using one of these strategies. Try it out with your students. Observe for evidence of productive struggle from Table 4.1. You can strengthen your ability to design instruction that promotes productive struggle and high-level learning by making a commitment to modify just one low-level math task each week.

Scaffolding Strategies to Turn Unproductive Struggle Into Productive Struggle

In SanGiovanni et al. (2020), the authors offered a wealth of tools for thinking about and supporting productive struggle, including a frame for noticing different types of struggle (p. 114) based on research by Warshauer (2014). We have added the examples in the second column (Table 4.2).

Because we want our students to develop the ability to push through struggle, when we notice that struggle is occurring, we must allow students to try resolving this struggle on their own. However, once struggle becomes unproductive, teacher scaffolding is needed. When a teacher has identified the type of struggle a student is experiencing, they are better positioned to choose a specific scaffolding strategy in the moment to help the learner turn unproductive struggle into productive struggle. SanGiovanni et al. (2020, pp. 115–137) offered this helpful list of teacher moves to support productive struggle (see Table 4.3). We have added the examples in the second column.

Table 4.2 • Five Types of Math Struggle

TYPE OF STRUGGLE	EXAMPLE
Unable to Get Started	Student doesn't understand what the problem is asking them to figure out.
Unable to Use a Process	Student doesn't know which operations might be used to solve the problem or can't represent the problem in a way that leads to a plan for solving it.
Unable to Calculate	Student understands the problem and has a plan for solving it but is unable to perform needed calculations.
Unable to Stay With a Task	Student experiences confusion and gives up before completing the problem.
Unable to Explain	Student solves the problem but cannot clearly explain their solution process.

Source: Adapted from SanGiovanni et al., 2020, p. 114.

 Available as a downloadable resource on the companion website.

Table 4.3 • Teacher Moves to Support Productive Struggle

STRUGGLE MOVE	EXAMPLE
Prepping the Task	Before students begin work on the task, encourage them to relish the opportunity to struggle because struggle leads to learning.
Catch and Release	As students work on the problem, you notice that a group of students is confused. You pause the class, pose an open-ended question related to the point of confusion, and orchestrate a brief class discussion to provide students with just enough information to continue thinking. You then send students back to work.
Referrals	When students become stuck, you refer them to a learning resource. These learning supports can include anchor charts, mathematical tools, and peers.
Metacognitive Questions	When students become stuck, you can ask a question that models the thinking you might do yourself if solving the problem. Use of metacognitive questions like the following can help students to develop the habit of asking themselves these questions as they engage in problem-solving. • What are some things you know about the problem? • What's a reasonable estimate for the answer? • How is this problem similar to other problems you've solved in the past? • How might you represent this problem? • What tools might you try to help you with this problem?
Remove the Numbers	When students are stuck, you can suggest removing the numbers from the problem or replacing them with easier numbers to first focus on the meaning of the problem.

Source: Adapted from SanGiovanni et al., 2020, pp. 15–137.

 Available as a downloadable resource on the companion website.

You can sharpen your eyes for productive and unproductive struggle and build your toolkit of moves for supporting productive struggle with these simple actions:

- Keep a list of the five types of struggle on a clipboard. As students engage in problem-solving, take brief notes about the types of struggle you observe.

- Choose one teacher struggle move that you want to strengthen. As you plan your math lessons, identify places to try this move. Afterward, reflect on the move's impact and your learning about scaffolding for productive struggle.

Sue's Math Story:

I (Sue) give my teaching candidates the Horse Problem (p. 102) during our first day of their Math Methods course to introduce them to the math habits.

We never come to consensus on the solution to this problem on the first day of class. The candidates continue to productively struggle with the Horse Problem across the week.

During our next class, candidates often share stories of passionate discussions about the problem over family dinners. Many candidates come ready to prove their solutions to others. As they debate the solution to the problem, candidates naturally begin to sketch out diagrams and tables on the classroom whiteboard to illustrate their thinking.

I never reveal the answer to the Horse Problem. Instead, we set aside a bit of time in each class session for candidates to share and discuss their thinking about the problem until the class reaches consensus on the solution. Frequently, candidates discover discussions of the Horse Problem online and explanations of the correct answer. Interestingly, however, this doesn't dissuade the class from continuing to discuss and debate the problem. It's as if our innate desire to make personal sense of a complex but interesting problem outweighs the need for an easy answer.

TOGETHER WITH YOUR TEACHING COMMUNITY

Since We Met Last (10 minutes)

In September, you and your colleagues gathered classroom data about how your students make sense of mathematical problems.

You might use the following protocol to share and reflect on this data with your math teaching community:

- Share your data with a colleague. (5 minutes)

- As a group, discuss what you learned from the data and how these insights can help you support your students' growth as mathematicians. (5 minutes)

Let's Do Some Math Together! (10 minutes)

In small groups, share your solutions to the Horse Problem. Does your group agree on the solution? If not, explain and justify your solutions to each other in an effort to reach a consensus.

Discuss:

- What math habits did you and your colleagues engage in as you debated the solution to the Horse Problem?

- What emotions surfaced for you and your colleagues as you wrestled with the problem on your own and as you discussed it with colleagues?

- How did you and others experience struggle as you worked through the problem? What specific things did you or your group members do that helped you push through this struggle?

Gonzalez (2023) said, "Too few people have ever had the opportunity to work on open-ended problems, make mathematical discoveries, and build mathematics for themselves" (p. 18). Is this true for you? When teachers have not had much experience tackling challenging math problems and experiencing productive struggle themselves, how can that impact their math instruction?

Building Our Expertise (30 minutes)

SanGiovanni et al. (2020) told us that "struggle is not something that happens by chance" (p. 4) and that educators need to intentionally and strategically incorporate opportunities for productive struggle into their design of learning experiences. They also told us that "productive struggle is more likely to be realized in individual classrooms when the whole school and all stakeholders value it" (p. 26).

Your school or team can promote productive struggle as a part of your mathematics program and help your students grow their "muscles" for perseverance by intentionally planning for struggle. Together with your colleagues, use these steps to plan a problem-solving lesson that allows your students to experience and learn about struggle and perseverance:

1. Select a thinking-rich math problem to use as the basis of a problem-solving lesson. (See the earlier discussion on Characteristics of Math Problems That Promote Struggle and Strategies for Modifying Math Problems to Promote Struggle.)

2. Try out the problem yourself, ideally with colleagues. Identify aspects of the problem that students may find challenging.

3. Plan ways you can support students during struggle without reducing the rigor of the task. (See Table 4.3.) Hint: It helps to have a couple of hip pocket questions related to the problem at the ready. You might write these questions on a sticky note to refer to as you check in with students while they work on the problem.

4. Plan how you will help students reflect on their struggle experience at the close of the lesson. You might use an open-ended question such as one of these for discussion or as an exit ticket:

   • How did you struggle today? Was it productive or unproductive struggle? How do you know?

(Continued)

(Continued)

- What did you do to persevere today? How did it go? What will you try tomorrow?
- What are you learning about productive struggle and perseverance? Why are these ideas important to know about?

How might you incorporate these perseverance planning steps into your weekly lesson planning routines?

Let's Try It (10 minutes)

Implement the problem-solving lesson that you and your colleagues planned together. As a teaching community, choose one of the data-collection focuses below. Bring this classroom data to next month's meeting to share and discuss.

Take time now to choose a date when you'll teach your problem-solving lesson and gather data from your chosen option. Record this task and deadline in your calendar or planner.

OPTION 1: PRODUCTIVE AND UNPRODUCTIVE STRUGGLE

As students work in partnerships or collaborative groups on the math problem, gather data about indicators of productive and unproductive struggle. You might print a copy of Table 4.1 and simply tally the behaviors that you observe. Or you could keep this listing of behaviors handy and take brief anecdotal notes about the behaviors that you see.

What are you learning about your students? What might their next steps in learning be related to productive struggle and perseverance?

OPTION 2: SUPPORT FOR PRODUCTIVE STRUGGLE

After teaching the lesson, jot down some brief notes about the impact of the struggle-support moves on your students' thinking and their efforts.

What worked well? What didn't go as you expected? What do you want to try next time?

**OPTION 3: STUDENT LEARNING RELATED
TO PRODUCTIVE STRUGGLE AND PERSEVERANCE**

Audio record the closing discussion from your problem-solving lesson or review students' exit tickets where they reflect on their learning related to productive struggle and perseverance.

What are you learning about your students? What might their next steps in learning be related to productive struggle and perseverance?

ADDITIONAL PROFESSIONAL LEARNING EXPERIENCES

This month's additional professional learning experiences relate to the mathematics content of whole-number and decimal place value. Our base-ten number system is an important mathematical structure that students must understand at a deep level by the time they leave elementary school. As members of your math teaching community have conversations about ways to develop this critical understanding across grade levels, teachers become better equipped to monitor student growth and to design learning experiences to deepen their students' understanding of place-value concepts.

Math Talks: Digit Place

This month's math talk is a highly engaging game that involves logical reasoning and gives students practice using place-value vocabulary to describe numbers.

1. Explain that you will think of a secret number and that the group will guess your number.

2. Share the rules for the secret number:

    • My secret number has ___ digits. (It's best to start with a two-digit number. After the group is comfortable with the game rules, you can play with a three-digit secret number.)

    • The secret number does not have a leading zero. (It will not start with a zero.)

(Continued)

(Continued)

- The secret number does not have any digits that repeat. (101 and 355 would not be allowable secret numbers.)

3. Draw a recording chart for guesses and clues on the board or chart paper:

GUESS	DIGITS THAT ARE CORRECT	DIGITS IN THE CORRECT PLACE

4. Call on someone to make the first guess. Record their guess and associated clues on the chart.

   Example: If the secret numbert is 125 and someone guesses 457, you would record the following:

GUESS	DIGITS THAT ARE CORRECT	DIGITS IN THE CORRECT PLACE
457	1	0

5. Take additional guesses and record clues.

GUESS	DIGITS THAT ARE CORRECT	DIGITS IN THE CORRECT PLACE
457	1	0
468	0	0
591	2	0

Stop periodically to ask the group to talk about what they know and don't yet know based on the clues. You might ask group members to turn and talk about what a good next guess might be and why or which numbers they feel they can eliminate and why. Or you might ask which can be eliminated as the secret number and how they know.

6. After playing several rounds, invite the group to discuss the learning value of the game. Ask "How did you engage in the math habits as we played this game?"

Note: Digit Place can be played with whole numbers or decimal numbers.

Discuss the experience:

- How did you engage in the math habits as you played this game?

- How does this game offer opportunities for productive struggle? What helped you to persevere in figuring out the secret number?

- How might you use this game in your classroom?

Inspired by the Digit Place activity in *About Teaching Mathematics* (Burns, 2022, p. 299)

Manipulatives and Models Matter: Base-Ten Blocks

We know that math manipulatives can help students tackle challenging problems, problems they might not otherwise be able to understand and solve (Hiebert et al., 1997; Huinker & Bill, 2017; Tapper, 2022). Manipulatives can serve as powerful scaffolds, temporary supports that allow students to access rigorous mathematics. They are also learning tools that all mathematicians, including adults, can draw on as needed to support our work with challenging mathematics.

Many of us did not have the opportunity to use math manipulatives as learning tools when we were in elementary school. And so, it's important for you and your math teaching community to regularly spend time exploring and developing meaning for the manipulatives that support your mathematics curriculum. As we deepen our own understanding of how these manipulatives represent the math we are teaching, we are better able to help our students use these manipulatives as tools to tackle challenging math problems.

As you engage in this activity with colleagues, notice how you are developing meaning for base-ten blocks as a mathematical tool and how tools can deepen understanding of the mathematical ideas they represent.

(Continued)

(Continued)

Try this experience with your math teaching community.

Represent several whole numbers using base-ten blocks, where the assigned value of the tiny cube is one or a unit. For the number 253, you need two "flat" hundred blocks, five "long" ten blocks, and three units.

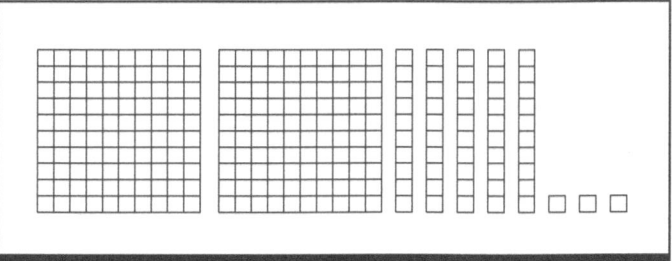

Now, change the definition of a unit. Assign the value of one whole to the "flat," thus, changing the value of the "long" to one tenth (0.1) and the value of the tiny cube to one hundredth (0.01). Using these values, represent several decimal numbers such as 1.42.

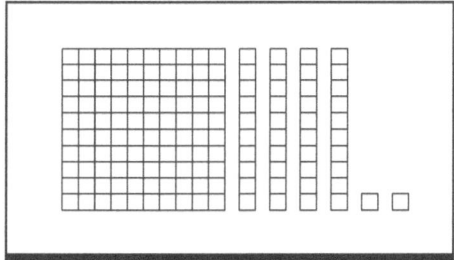

Discuss how these representations of whole numbers and decimal numbers are similar and different. What big math ideas lie behind representation of whole numbers and decimal numbers with base-ten blocks? Why are these ideas important?

Because teachers have few opportunities to think about the mathematics curriculum beyond their grade level, we sometimes teach place value as a set of isolated skills, missing opportunities to help students understand and use the underlying base-ten place-value structure as a way of thinking about numbers and our number system. When students who have used base-ten blocks to represent whole numbers learn how to use this same tool to represent decimal numbers, they analyze and apply the mathematical structure of our place-value system. In doing so, they deepen their understanding of base-ten place value as they build understanding of whole numbers and decimal numbers.

REFLECT

- How do base-ten blocks help you to think about decimal numbers?

- How might you use mathematical tools to deepen your student's understanding of the math they are learning?

- How can mathematical tools support productive struggle and perseverance?

Game Time

This month's game supports understanding of whole-number and decimal place value. It is a favorite with students. It's simple to learn and easily adaptable to multiple grade levels and a variety of math content.

RACE FOR 100

Materials:

2 players

Two dice for each pair of students

Base-ten blocks

Challenge: Be the first player to get 100.

Game directions:

1. Partners take turns rolling both dice.

2. Take the number of base-ten units indicated on each die.

3. Once a player has 10 units, they must trade them in for one ten, a "long."

4. The first player to accumulate 10 "longs" trades them in for one hundred, a "flat." They are the winner!

(Continued)

(Continued)

Depending on your specific student learning focus, you might:

- Provide a sentence stem to support students' use of math language:

- I took ___ ones. Now I have ___ tens and ___ ones. My total is ___.

- Have students record each move with an equation to practice representing multi-digit numbers in expanded and standard form:

$$4 \text{ tens} + 8 \text{ ones} = 48 \text{ or}$$

$$40 + 8 = 48$$

Increase the challenge and the learning value of the game by having students write about their learning. You might use a prompt like this:

> Use at least three words from the word bank to write about your learning.
>
> Word bank: ones, tens, hundreds, digit, place, value, total, standard form, expanded form

Variations

- Play **Race for 100** using a hundreds chart and marker rather than base-ten blocks.

- **Race for $1.00** follows the same rules but uses pennies, dimes, and a dollar rather than base-ten blocks.

- **Race for 1** focuses on decimal concepts. It follows the same rules as above. Base-ten blocks are used with the "flat" having a value of one. The dice roll indicates the number of hundredths that a player should take.

- **Race for 0** can use either base-ten blocks or money. Play begins with 100 or $1.00 and the dice roll tells how many ones or pennies to take away. The winner is the first person to reach 0. Race for 0 can also be played with decimal values.

Discuss the experience:

- What math habits did you engage in as you played this game?

- How can games like this help students learn to confidently use manipulatives as learning tools? How does the ability to use mathematical tools support productive struggle and perseverance?

Inspired by the *Roll for $1.00* game in *Math Games for Number and Operations and Algebraic Thinking* (Petersen, 2022)

online resources   Available as a downloadable resource on the companion website.

IN YOUR CLASSROOM

Classroom Story

Recently, Ms. Soliz noticed that when her students were working on challenging math problems, some students stayed engaged with a complex problem longer, trying out multiple strategies and tools, talking about their math thinking, and asking their classmates questions. Other students gave up when they reached an impasse, a point where they were confused or mentally exhausted. When this happened, Ms. Soliz would ask a question to help the students think about the mathematics they were struggling with. Although this scaffolding technique often allowed the student to reengage in the problem, Ms. Soliz knew it was important for students to develop agency for their own learning. She wanted each of her students to have the internal resources needed to persevere with challenging math.

Ms. Soliz decided to introduce her students to the concept of perseverance in the context of a problem-solving lesson. Understanding what it means to persevere and why perseverance is important would help students to see perseverance as a skill set that can be learned. Then, over time, Ms. Soliz would help each of her students build personal toolkits of perseverance strategies.

The class had been learning about decimal numbers. Yesterday, they had investigated all the different ways they could represent 1.24 with base-ten blocks. Today, Ms. Soliz wanted to help her students make connections between what they were learning about decimal place value to what they already knew about money. She gave them this problem to solve.

- What are all the different amounts of money you could make with 17 coins (dimes and pennies only)?
- List these amounts in order.
- Write about the patterns you see in your list.

Based on the Eight-Coin Problem (De Francisco & Burns, 2002, pp. 110–118)

Ms. Soliz asked the class to identify the largest and smallest amounts of money that could be made. The class agreed that $1.70 was the largest possible amount and that $0.17 was the smallest amount. Ms. Soliz then asked students to explore the problem independently.

As she watched the students at work, Ms. Soliz took the following notes about struggle and perseverance that she would use when the class debriefed from the problem-solving experience.

Examples of struggle:

- Not understanding the problem
- Found several answers but then became stuck

- Difficulty keeping track of mental calculations
- Worn out—exhausted from the mental effort of calculating and recording
- Not able to determine whether all answers had been found
- Not able to explain the patterns in the answers

Examples of perseverance strategies:

- Asking questions
- Using coins to find answers
- Looking for patterns and using these patterns to find missing answers
- Experimenting with different ways of keeping track of the number of each type of coin and the amount of money each coin combination represented
- Working together, talking, and sharing ideas

During the class discussion at the end of the lesson, Ms. Soliz had several students share their solutions to the problem using the document camera. The class discussed the patterns that they saw in the list of decimal numbers and conjectured about why these patterns occurred.

Then Ms. Soliz wrote the word "perseverance" on the board. She explained the idea of perseverance by giving examples from the notes she had taken. Ms. Soliz asked students to consider why perseverance is important in learning. As students shared, Ms. Soliz made connections for the class to the examples of struggle that she had observed.

In their exit tickets, Ms. Soliz asked students to respond to this prompt: Today I persevered by ____. One student called Ms. Soliz over to his desk and whispered that he hadn't persevered today. Ms. Soliz responded, "You can write that. And tomorrow, we'll keep working on perseverance."

Anchor Lesson

Teaching Students How to Persevere

WHAT?

This lesson will help students begin to recognize situations that call for perseverance and consciously draw on internal and external resources to persevere with challenging mathematics.

YOU NEED:

Challenging math task

Chart paper

Markers

HOW?

1. Prepare an anchor chart with the word "perseverance" at the top. You will use this at the end of the lesson.

2. Engage students in a challenging mathematics task. You might choose a task from your mathematics curriculum, or try out one of these problems:

   - Kindergarten: How many different ways can you make 10?

   - First grade: How many different ways can you make 20¢?

   - Second and third grades: How many numbers can you make using the digits 1, 2, 3, and 4. You may only use each digit once in each number that you make.

   - Fourth and fifth grades: What might the missing digits be in this decimal addition equation? $3.\_1 + \_.47 + 0.\_ = 8.68$

3. As students work on the problem, observe and listen carefully for signs of productive and unproductive struggle. You might use ideas from Figure 4.3 to help students persevere.

4. At the end of the lesson, return to the anchor chart. Give examples of perseverance behaviors from your observations of your students at work. Invite students to share other examples. Together with students, record strategies for persevering on the anchor chart. Ask students to talk about why perseverance is important.

REFLECT

Perseverance can be a challenge to experience and teach. Students need safe spaces and opportunities to persevere while not becoming overly frustrated. We can intentionally generate experiences that force students to persevere and make time to reflect on their behaviors and attitudes needed to use this important skill.

Use the following discussion questions to reflect with students:

- What are some ways that you persevered today?

- How can you persevere in the future?

LET'S DO SOME MATH ACROSS OUR SCHOOL!

Burns (2015b) used the Horse Problem with students in second grade and up. This problem can allow students to experience and talk about productive struggle. You might simplify the numbers for younger students:

A person bought a horse for $1 and sold it for $2.

This same person then bought the horse back for $3 and sold it again for $4.

Did the person make money or lose money? How much did they make or lose?

You can also modify this task to make it more relevant to your students' experiences. You might, for instance, revise the story to involve buying and selling a bicycle or another object that students can easily identify with.

As your students make sense of and work through this problem, you'll want to engage them in acting out the problem and using manipulative models to represent the quantities of money exchanged. Consider having students record their math thinking with words, numbers, and pictures. If all grade levels display students' solutions in the hallways outside of classrooms, students, educators, and parents will be able to see and celebrate all of the mathematical brilliance that exists within your math learning community.

You can read more about the Horse Problem in Burns's blog post (2015b): https://bit .ly/4cIPKR7

TEACHING MOVE: FOCUSING AND FUNNELING QUESTIONS

The mathematics instruction many of us experienced as children involved the teacher asking a series of questions designed to "funnel" our thinking down the teacher's own thinking pathway toward a correct answer. **Funneling questions** such as "Which digit is in the tenths place?" or "How do we write 4,856 in expanded form?" offer students limited opportunities for sense-making because the goal is for students to identify the answer to a single problem rather than to strengthen mathematical understanding and critical thinking.

If we want our students to value productive struggle as a means of growing their mathematical understandings and skills, then our questions need to help them learn to look for and focus on important mathematical ideas. **Focusing questions** such as "What do you know about 0.3?" or "When you look at the counting numbers from 100 to 200, what patterns do you notice?" help students build the habit of figuring things out for themselves.

According to education researchers Hattie et al. (2017, p. 89), the teaching practice of skillful questioning has an effect size of 0.48 on student learning, placing it within the category of practices known to have a high impact on students' academic success.

APPLY

Audio record yourself facilitating a class discussion. Afterward, listen to the questions you asked during this discussion. Notice the focusing and funneling questions that you asked during this discussion. How did they impact your students' thinking and the learning that occurred during the discussion?

Prior to your next class discussion, preplan a couple of focusing questions that you will pose. Notice what happens when you ask these questions.

REFLECT

How can focusing questions support productive struggle and perseverance?

CLASSROOM ROUTINE: SYN-NAPS

Syn-Naps is a term coined by the neurologist–educator team Willis and Willis (2020) that refers to a quick processing activity that allows students' brains to shift gears and integrate new learning into existing schema.

Here are some examples of syn-naps:

- Stop and jot (Take 2 minutes to jot down an idea you want to remember.)

- Stop and draw (Take 2 minutes to sketch one way you might represent this math idea with a diagram.)

- Stand and share (Stand up. Take six giant steps. Connect with a person near you and tell them about your math thinking.)

Willis and Willis explained how syn-naps support learning:

> During syn-naps, the newly learned material has the opportunity to pass beyond the temporary short-term and working memory while students replenish their supply of neurotransmitters and the amygdala cools down. After just a few minutes, students' brains are refreshed. They are ready to return to the next learning activity with receptive amygdala and a full supply of neurotransmitters. (2020, pp. 62–63)

When you notice that student struggle with a challenging math problem is intense, pause the class for a syn-naps, perhaps followed by a brief discussion of perseverance strategies they are trying and how these strategies are working. You can also encourage individual students to take a quick syn-naps or brain break when you notice that their struggle is progressing from productive to unproductive.

STATION ACTIVITY: KENKEN PUZZLES

Professional athletes know they must work hard to strengthen their athletic performance. They struggle both physically and emotionally with this work, and yet they find the opportunity to stretch beyond their current abilities to be exhilarating. What if students saw challenging mathematics in this same way, an exciting opportunity to stretch their cognitive abilities, a chance to see what they are capable of and then push beyond their current edge of understanding? You can develop the idea that challenging math can also be fun with a standing classroom learning station stocked with math puzzles, number tricks, and games. Here's a puzzle idea to get you started.

KenKen puzzles engage students in logical thinking along with practice of basic number facts. In this partially completed KenKen puzzle, you must place the numbers 1, 2, and 3 in individual boxes so that each number appears once and only once in each row and column. The clues in the top left corners of some boxes tell the sum of the numbers in the dark outlined shape where the clue appears.

Can you complete the puzzle?

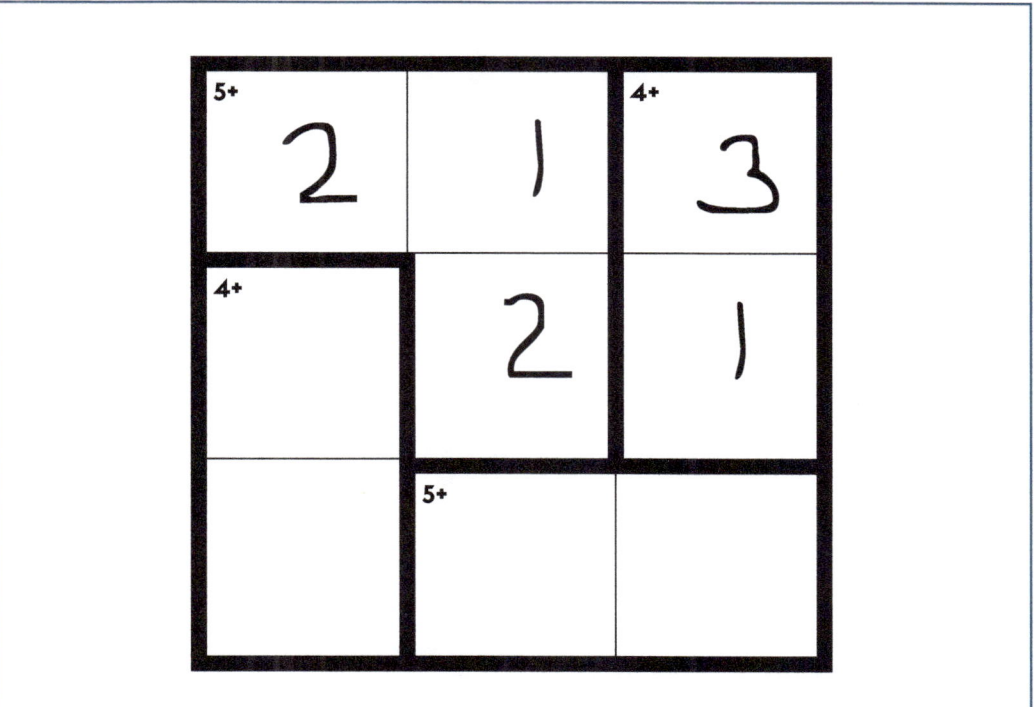

(Continued)

(Continued)

You can learn more about KenKen puzzles in Burns's blog post (2015c; https://bit .ly/3wGz8OC)

For younger students, consider using tangram puzzles. Mathigon.org is a free website that offers a wealth of hands-on experiences for students. Students can select a specific tangram template and fill in the pieces to complete the puzzle. (https://mathigon.org/ tangram)

 Available as a downloadable resource on the companion website.

LITERATURE CONNECTIONS

***Lia & Luís: Puzzled!* by Crespo** (2023) (Grades K–2)

Twins Lia and Luís receive a mysterious package from their grandmother in Brazil. Inside is a puzzle with a secret message, but putting the puzzle together turns out to be a challenge involving geometry concepts and spatial relationships. Lia and Luís persevere, however, solving the puzzle and revealing the secret message and a surprise.

Activity: What are some things Lia and Luís do to persevere in solving the puzzle? Work with a friend to solve a jigsaw puzzle or another kind of puzzle. What did you and your friend do to persevere? How did you use math ideas to help solve the puzzle?

***Grace Hopper: Queen of Computer Code* by Wallmark** (2017) (Grades 3–5)

Computer scientist Grace Hopper once said, "I have insatiable curiosity. It's solving problems. Every time you solve a problem, another one shows up behind it. That's the challenge." Grace loved solving problems involving mathematics. Because of her work with computer programming for the Navy, she earned the Presidential Medal of Freedom.

Activity: Some people think a problem is a bad thing, but Grace loved problems. Why do you think she enjoyed solving math problems? What kinds of math problems do you enjoy solving?

SPOTLIGHT ON BRAIN SCIENCE

It's important for students to know that we all get anxious from time to time, and that anxiety affects our ability to learn. Students also need to know that we can learn to monitor and regulate our emotions. When we recognize feelings of anxiety, we can take steps to lessen our stress so that we can re-engage in learning and persevere with challenging tasks (Swanson, 2013; Willis & Willis, 2020).

You can share this information with students in a mini lesson. You might also use this reproducible for a station activity or to send home along with the Family Newsletter to help families talk about these important ideas.

KNOWING HOW OUR BRAINS WORK CAN HELP US PERSEVERE.

The amygdala is a part of our brains that plays an important role in productive struggle and perseverance. When we struggle productively on a challenging math problem, our neurons or brain cells link up and we learn a lot. But when we are stressed, our amygdala sends out signals in our brain that keep us from learning. This is sometimes called "the run response" because it can make us feel like running away from learning.

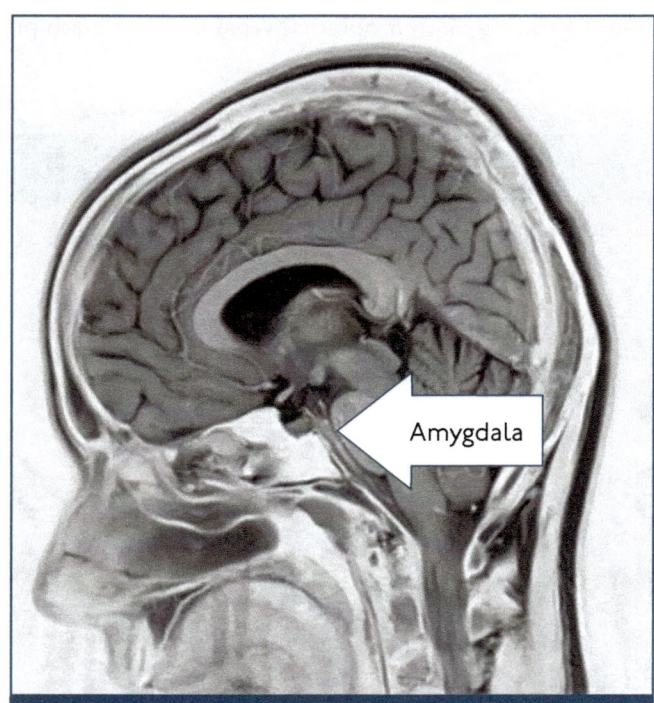

Image source: lstock.com/springsky

As you work on challenging math, you can learn to notice when your amygdala is stressed and then take steps to help your amygdala calm down.

1. As you work on a challenging math problem, stop every once in a while, and think about how you are feeling. You might use emojis to help you decide.

| I am struggling productively. | This is pretty challenging. | I am stuck or frustrated. |

Images source: Istock.com/mooikunst

2. If you are feeling stuck or frustrated, use a stress-busting strategy such as the following to help your amygdala calm down:
    *   Take several slow deep breaths.
    *   Tell yourself, "I can do this!" Visualize yourself persevering and succeeding.
    *   Take a brain break. Do something different for a couple of minutes. Get a drink of water and, if possible, move around a bit.

After you finish the math problem, stop and think about what you tried and how it worked. Make a plan for what you will try the next time you have the chance to work on a challenging math problem. Before you know it, you'll be an expert at persevering with challenging math.

Choose one of these ways to think more about this important information:

1. Write down why these ideas are important to you and how you want to put them to work in your life.

2. Talk to a friend about these ideas. Decide together how you will put them to work.

3. Share these ideas with someone at home. Tell them why these ideas are important to you and how you plan to use them.

online resources ➤ Available as a downloadable resource on the companion website.

SPOTLIGHT ON EQUITY: ALL STUDENTS DESERVE TO STRUGGLE

Research has shown that minoritized students, students with a first language other than English, and students whose family income falls before the federally established line for poverty sometimes receive less rigorous instruction and therefore have fewer opportunities for productive struggle (Hammond, 2014; Wedekind & Thompson, 2020). Educators want all students to become strong mathematical thinkers and problem solvers, but persisting misconceptions about the role of struggle in learning sometimes lead teachers to reduce rigor rather than maintaining high cognitive demand by scaffolding up to allow students access to rigorous learning (Zavala & Aguirre, 2024). This reality impacts students' immediate learning and their readiness for future learning challenges. Kobett and Karp reminded us that all students need regular opportunities to struggle with challenging mathematics tasks:

> After students carry out these tasks over time, they become more ready and able to tackle a challenge. They know that in giving them such tasks, you as their teacher have a high regard for their ability, which in turn increases their self-efficacy. They learn to be unafraid of cognitive dissonance, they begin to love puzzles and counterintuitive situations, and they begin to trust themselves to persevere. They feel more confident leveraging their own strengths, even when the pathway to success may not be clear. For all of these reasons, rigorous mathematics using high-cognitive tasks is not just for some—it is for all. (2020, p. 124)

SanGiovanni et al. (2020) reiterated this important message: "All students deserve the right to struggle. It cannot be reserved for the students who appear to need extra challenge or enrichment" (p. 3). This belief that all students have the right to struggle so that they can gain access to important mathematical understandings is foundational to equitable math classrooms. It has important implications for our work with students who receive special education services (Lambert, 2024), as well as with other students who have been historically marginalized due to race, language, and socioeconomic status (Boaler, 2019).

We position our students for success both now and in the future when we believe in their capacity to tackle and persevere with challenging mathematics and then design learning experiences and supports that reflect this belief.

TRY IT

Think of a student who currently resists struggle. Set aside a brief time to conference with this student about how struggle supports learning and emphasize that persevering through struggle is a skill that can be learned. Support the student in setting a small goal related to their willingness to engage in struggle. Together create a plan for specific things the student and you will each do to work toward achieving this goal. Schedule a time to meet again to monitor progress.

MATHEMATICAL ME: STUDENT JOURNAL AND PORTFOLIO

Student Journals:

Have students respond to the following questions:

- What is productive struggle, and why is it important?

- What are strategies I can use to persevere with challenging problems?

Older students can write their responses in their Mathematical Me journals. Younger students can draw a picture showing their response, or you might gather responses during a class discussion or quick one-on-one interviews.

Review students' responses to monitor students' beliefs about productive struggle and their strategies for persevering with challenging math. You might summarize this data by tallying or graphing the different ideas students offer.

Be sure to share and discuss this data with your class. Ask, "What does this data tell us?" and "What are some things we can do to continue growing as mathematicians?"

(Continued)

(Continued)

Student Portfolios:

Have students choose a piece of work where they persevered to solve the math problem. On a sticky note, have them write the date and complete this sentence frame:

Source: Istock.com/Thammask-Chuenchom

FAMILY NEWSLETTER: WE LOVE CHALLENGING MATH PROBLEMS!

WHAT WE'RE LEARNING

This month we're learning that challenge and struggle are essential to learning and that we can all get better at persevering in our learning with practice and support.

WHY IT'S IMPORTANT

Struggle can be both good and bad. When we persevere with challenging math tasks, we grow new understandings and skills. But when we become stuck or frustrated with math tasks, learning doesn't occur. We feel discouraged and sometimes give up. We all need a toolkit of strategies to push through struggle into learning.

The ability to persevere through challenge is both a learning skill and a life skill. Like all skills, it can be strengthened with practice, especially when a child has a caring mentor like you on their side to encourage and support them in learning how to persevere.

HOW YOU CAN HELP

You play an important role in helping your child develop the habit of persevering in real life and school. As your child tackles daily tasks that feel challenging (e.g., cleaning their room or saving money for a special purchase), you can help them set goals and take pride in small steps toward these goals. If your child has a favorite sport or a hobby, you can help them see how their practice of this activity can feel like struggle but is essential to their developing expertise. You can also help your child make connections from real-life experiences with struggle and perseverance to their learning work at school.

You might ask questions like these:

- Think of a time when you worked really hard to learn something new. How did that feel at first? How does it feel now?

- When something is challenging, what are some things you can do to make sure you stick with this work until you reach your goals?

online resources ↖ Available as a downloadable resource on the companion website.

CHECKING IN ON OUR LEARNING: WE LOVE CHALLENGING PROBLEMS!

- What is productive struggle, and why is it important?
- What are strategies I can use to persevere with challenging problems?

This month's learning focus, embracing challenging math problems, is central to what it means for students and educators to be math learners and mathematicians. When we enthusiastically engage in struggle and persevere in making sense of and solving complex math problems, we live rich mathematical lives. In a powerful math community, educators are skilled at designing instruction and providing learning supports that promote productive struggle. They also take responsibility for teaching students about the value of struggle and strategies for persevering with challenging mathematics.

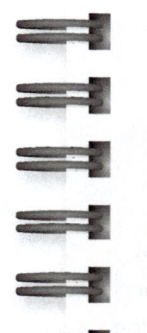

MATHEMATICAL ME: EDUCATOR JOURNAL

- How has this month's learning focus supported your students' mathematical growth?
- How has this month's learning focus supported your growth as a math teacher?
- How has this month's learning focus supported your school's growth as a powerful math community?

LOOKING AHEAD

In November, we will explore the idea of mistakes as important stepping stones to learning. We'll consider feelings associated with making mistakes and the learning environment that must be in place for students to see mistakes as valuable.

Questions for Reflection

- How do your students react when they realize they've made a mistake in math?
- What is a mistake you've made in the past that turned out to be a good thing?

CHAPTER FIVE

November
We Learn From Math Mistakes!

ESSENTIAL QUESTIONS

- How do mistakes help me learn math?
- What should I say or do when I make a math mistake?
- What should I say or do when someone else makes a math mistake?

THIS MONTH'S FOCUS

In a powerful math community, students, educators, and leaders understand that mistakes are a key part of learning. They see each new problem as an opportunity to take a risk because they understand that learning is messy. They rely on each other to make sense of math, reason about their strategies and answers, and use mistakes to create a deep understanding of math content.

Teachers and leaders help students understand that math learning involves a never-ending cycle of making sense of problems, trying out new strategies, analyzing results, and deciding on the next steps.

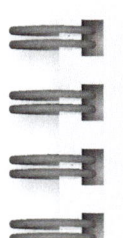

MATHEMATICAL ME: EDUCATOR JOURNAL

Take five minutes to respond to the essential questions. You will have a chance to reflect on these same questions and to have students respond to these questions at the end of the month.

ON YOUR OWN

LET'S DO SOME MATH!
DESIGN YOUR OWN T-SHIRT QUILT

We use mathematics in a variety of real-life activities, often without thinking about the fact that we're doing math. Perhaps creative work like quilt design doesn't feel like mathematics because we're not preoccupied with "finding the right answer" or fearful of "making a mistake" in our design. Instead, we're simply exploring what's possible and what's pleasing and using mathematics to achieve this goal.

A t-shirt quilt, like the one above, allows its creator to capture special memories in a practical and beautiful object. A lot of math thinking goes into the design and construction of the quilt. A full-size t-shirt quilt typically measures 81 × 96 inches when complete.

Knowing you need a border, room for t-shirts, and filler fabrics, create a blueprint for your quilt. Make sure that it has some symmetry. A suggested design follows.

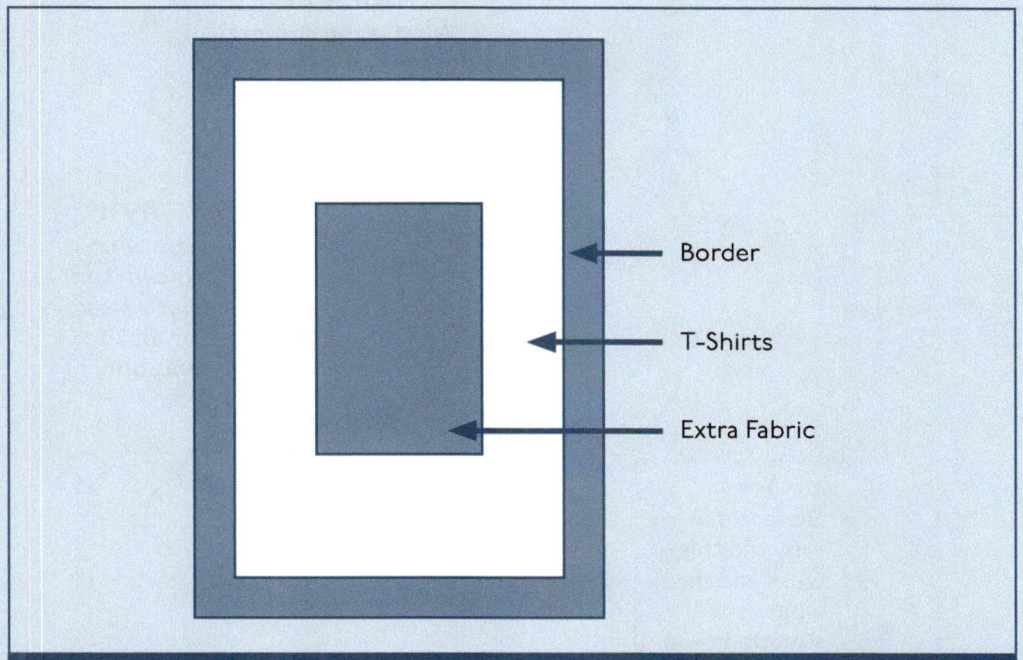

Consider how you engaged in the habits of mathematically powerful people as you designed your quilt.

 Available as a downloadable resource on the companion website.

Why This Focus

How do we define mistakes? The word **"mistake"** has a negative connotation. I (Holly) find that it doesn't serve me when talking about math learning. Children often hear the word "mistake" and think that it's something they should feel ashamed of. In the context of the classroom, I like to think of students' missteps as "learning" rather than as "making mistakes." If we see learning as a continuous cycle of making meaning, trying something out, determining if the result fits the situation, and either coming to a conclusion or starting the cycle over, then making a mistake is really *learning*. Consider the learning cycle shown in Figure 5.1.

Figure 5.1 • The Learning Cycle

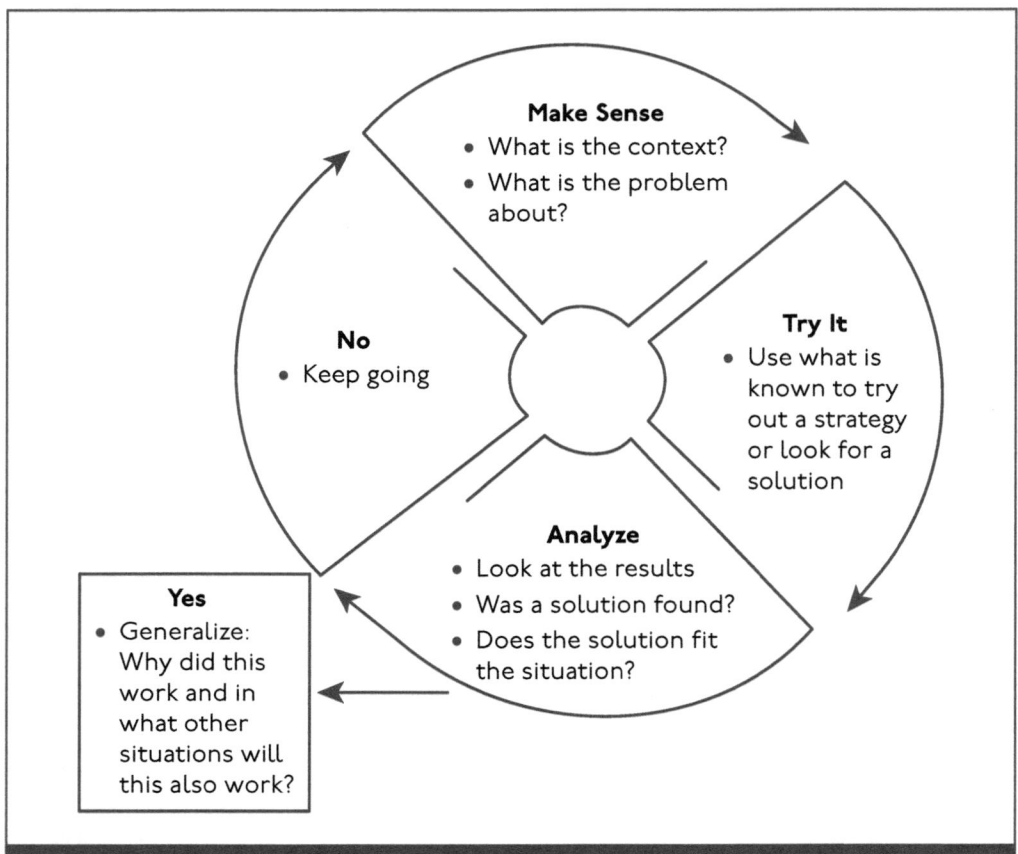

Let's imagine this learning cycle in a real-life context.

Holly's Math Story:

I grew up baking, and it all started with my easy bake oven. I watched patiently as a light bulb baked doll-size cakes and cookies. Eventually, I graduated to a real oven and spent most of my time making cookies, cakes, and muffins. Into adulthood, I wanted to master bread making. Just as it had been with cake and cookie recipes, I expected that by adding all the ingredients into a bowl and mixing, I would produce a fluffy, soft, and delicious bread. To my surprise, this wasn't the case, and instead I wound up with a dry, hard, crumbly bread that was borderline inedible. In my first bread-making experience, I made it through the first three steps in the learning cycle. At that point, I needed to start the cycle over again to produce a successful result.

Years later an unexpected situation led me to try bread making again. I was teaching fifth grade, and part of our science curriculum involved working with various foods to determine sugar content by using yeast and warm water. Students learned yeast needs warm water and sugar present to grow. Foods with high sugar content placed in a plastic bag, when mixed with

> If we see learning as a continuous cycle of making meaning, then making a mistake is really *learning*.

yeast, would cause the bag to balloon quickly. Observing my students carry out this experiment, I had my "aha" moment. I wondered, if I applied the same principles to making bread, would I be able to make edible bread?

After applying my new learning about adding yeast and sugar to warm water, I'm happy to report that I now make delicious garlic and cheddar bread that my entire family requests frequently.

If we can look at mistakes as simply a natural part of the learning cycle, we begin to see that these moments are necessary to learning and teaching for understanding. I may have followed the recipe the first time, but I didn't understand why putting the ingredients together in a particular order would produce quality bread.

As we think of mathematics learning in the classroom, we can apply these principles in the same way. We ask students to read and make sense of problems. Then, they use what they know to try the problem. Often, classes in which students arrive at different answers make for great jumping-off points for analysis and trying again. The "mistakes" they make lead to an understanding of the concepts at a deeper level.

Tapping Into Your Experience

As we consider the role mistakes play in learning, it is helpful to consider our own learning experiences.

Think back to a time when you learned something well. Maybe you're a good cook who spent years perfecting your methods and recipes. Perhaps you learned to build cutting boards and bookshelves that your family and friends proudly display in their homes.

What were your experiences like in learning that new skill?

Holly's example:

WHAT DID YOU LEARN?	EXPERIENCES:
Learning how to cook	I was exposed to it when I was younger.
	I enjoyed eating food.
	When I made something that someone enjoyed, I felt accomplished and satisfied.
	I was able to try new things.
	Sometimes, my methods or recipes weren't great, so I read about why.
	I could access the Internet to read about ways to make my breads softer or rise better.

It can be hard to remember what it was like when we didn't know something that we now do all the time. How did I learn to feed myself? How did I learn letter sounds so I could read? One would have to expect that along the way there were many times in learning to eat when we made a mess. Or, in learning to read, we misread words and got stuck.

Just like these experiences, when we learn math, we make mistakes. If the conditions are right, if we have meaningful experiences, or if someone encourages us to analyze why what we tried didn't work, we can successfully learn something new. Here are some other conditions to consider that support learning. What else might you add to the list below?

CONDITIONS FOR LEARNING

- Having a support person
- Engaging in multiple meaningful experiences
- Time to reflect after each attempt
- Patience from those who support us
- Interest in the topic
- Small successes
- Resources
- A community of learners

The Role of Shame in Framing Mistakes

As we explore our mindset and beliefs about mistakes in learning, we need to consider the role of **shame**. Researcher and author Brené Brown (2013, para. 2) defined shame as "the intensely painful feeling or experience of believing that we are flawed and therefore unworthy of love and belonging—something we've experienced, done, or failed to do makes us unworthy of connection."

- When have you experienced shame as a learner?
- How did that experience shape your learning experiences?

Shame is an emotion I (Holly) experienced as a young student sitting in a classroom. If my answer was wrong, the teacher's response was simply "no," accompanied by a look of disappointment. The teacher would move on to another student, hoping to hear the correct answer, while I was left feeling deflated and confused, not really knowing where I went wrong. In her book *Math Therapy*, Vanessa Vakharia wrote:

> As a result of our natural inclination toward negativity, our math stories are often filled with storylines that put us on high alert.

Someone might have it in their head that when there's a risk of getting a math problem wrong, that they're likely to experience shame, panic and confusion. The negativity bias makes them overly concerned about making mistakes, potentially hindering their willingness to take risks in learning. (2025, p. 189)

I learned that to be in math class means to be quiet unless you can guarantee your answer is correct.

As we learn more about how the brain and emotions affect learning, it's worthwhile to reexamine the teaching moves we use to respond to mistakes. How can we proactively respond to mistakes in ways that don't trigger shame, creating feelings of isolation and unworthiness?

Recall an experience when you felt shame after giving a wrong answer. What type of response to your mistake from those around you might have changed how you felt at that moment?

Once we believe mistakes, incorrect answers, and partial thinking are all necessary for learning, we can quickly move past negative, shameful feelings. As educators, we start to see mistakes as a welcome and joyful experience that indicates students are engaging in math, learning, and using higher order thinking skills to formulate their ideas and solutions. SanGiovanni et al. (2020) pointed out that struggle is essential to living and learning: "The lessons students learn, and the skills they acquire by successfully working through struggle in school, will set them up to be more successful in situations when they face struggle in the future" (p. 16).

It can be challenging to shift our teaching practice. We often emulate what we experienced as students in school. Maya Angelou reminded us, "I did then what I knew how to do. Now that I know better, I do better" (quoted in D. Smith, 2021). When we recognize that we have used ineffective or even harmful teaching practices in the past, we too need to move past shame and into learning. Consider these teaching moves to help students reframe mistakes as learning opportunities:

- Listen to all responses without judgment. With both correct and incorrect answers alike, keep a neutral face and tone.

- Record all answers on the board, even ones you know are incorrect.

- Try using language that opens up discussion. For example: "I see there are a couple of solutions to this problem. How can we figure out which one is correct?" "Who would like to defend one of the solutions on the board while the rest of us listen and think about how this idea is alike or different from what we thought?"

> Once we believe mistakes, incorrect answers, and partial thinking are all necessary for learning, we can quickly move past negative, shameful feelings. As educators, we start to see mistakes as a welcome and joyful experience that indicates students are engaging in math, learning, and using higher order thinking skills to formulate their ideas and solutions.

When we transform our classrooms into places where students aren't afraid to take risks and share what they have tried, students feel empowered to respectfully debate answers and come to conclusions based on the collective knowledge they share. At the same time, teachers gain a wealth of knowledge about their student's understanding and can use that information to make instructional decisions.

Sharing Mathematical Authority

In his book *The Imperfect and Unfinished Math Teacher*, Orton (2022) reminded us, "When we practice the Art of Not Knowing in our classrooms, we position ourselves to share authority with our students because they see us as learners with them—instead of seeing us as someone who is holding all the answers." (pp. 2–3).

- Consider a time when you made a mistake in front of your students. Maybe you didn't know the answer to their question. What feelings does this bring up for you?

- If we worried less about not knowing and focused our attention on listening and learning in the moment, how might the experience of learning mathematics change for students?

Meyer (2023) has highlighted the work of math teachers nationwide. In his article "AI Does Not Fit the Shape of Schooling," he discussed teacher practices that accentuate the humanness of teaching and learning. He described a classroom in which the teacher skillfully records a student's incorrect answer on the board, trusting that the class will respectfully consider this answer, use it to deepen their understanding, and determine the correct answer together (https://bit.ly/3SYrbvu).

- Why might the practice of accepting all responses and trusting students to come to a consensus on the correct answer be important?

- As an experiment, try this out in your own classroom. Record all student responses without judgment. Follow up by asking students, "How can we tell which one of these responses is correct?" Notice what happens.

In previous chapters, we explored the essential skills of students being able to persevere, productively struggle, and make their own sense of mathematics. When our students know that we care about what they say, regardless of being right or wrong, they are better positioned to use those critical thinking skills necessary for a deep understanding of math. Part of the power in sharing mathematical authority is the positive emotional states our students experience. Our classrooms become places of joy and satisfaction, according to McConchie and Jensen (2020), fostering a positive climate for learning. They wrote that "satisfaction comes when students feel connected to others, have control over their day, see progress in what they do, and have a sense of purpose or meaning" (2020, "Consider Emotions," para. 2). Sharing mathematical authority means we activate positive emotional states, making room in our students' brains to actively engage in sense-making, perseverance, and productive struggle.

TOGETHER WITH YOUR TEACHING COMMUNITY

Since We Met Last (10 minutes)

In October, you and your colleagues gathered some classroom data about perseverance strategies your students are currently using.

You might use the following protocol to share and reflect on this data with your math teaching community:

- Share your data with a colleague. (5 minutes)

- As a group, discuss what you learned from the data and how these insights can help you to support your students' growth as mathematicians. (5 minutes)

Let's Do Some Math Together! (10 minutes)

Share your quilt design with your partner. Discuss the mathematics you used in thinking about your quilt design.

Discuss:

1. How did you experience the learning cycle while you worked on the problem?

2. The creation of a quilt might be thought of as a metaphor for the learning process. How do you see these two ideas as similar?

3. How did this experience help you to think about the role of "mistakes" in learning?

Building Our Expertise (30 minutes)

It can be challenging to focus on mistakes in the mathematics classroom. If your experiences as a student were anything like mine (Holly), the goal in math was to get to the correct answer, and quickly. You may question that the time you set aside to focus on mistakes may not be worth the effort. What might be getting in the way of this essential practice?

Spend five minutes individually journaling about the following questions:

- What concerns you most about allowing students to share work that is incorrect or confusing?

- What might be the long-term benefits of allowing students to share mistakes and thinking that isn't yet clear?

Talk in partnerships about your thinking.

As a whole group, share and record the group's ideas on two charts, which might look like these:

1. Concerns about Sharing Incorrect Work
   - Students will gain misconceptions
   - It takes too much time
   - Students will pick up bad math habits
   - Students will feel bad

2. Long-Term Benefits of Sharing Mistakes
   - Students learn everyone makes mistakes
   - Students learn what to do when they make a mistake
   - Students learn the skills of analyzing, justifying, reasoning, and communicating
   - Students may understand concepts deeper since they know what it *doesn't* look like

Next, consider the concerns that have been listed. Break into small groups and assign each concern to a different group. In your small groups, brainstorm solutions to support your colleagues in addressing this concern.

Let's Try It (10 minutes)

As a teaching community, choose one of the data collection focuses that follow. Bring this classroom data to next month's meeting to share and discuss.

Take time now to choose a date when you'll gather this data. Record this task in your calendar or planner.

OPTION 1: WATCH CHILDREN'S FACIAL EXPRESSIONS WHEN THEY MAKE A MISTAKE.

Notice three students' reactions to their mistakes and record what you observe.

OPTION 2: JOURNAL ON INDEX CARD.

Ask students to think about the question "How do you feel when you make a mistake?" Allow students some time to think followed by a brief pair–share. Ask students to respond to this sentence stem on an index card or lined paper: **When I make a mistake, I feel _____.**

Save a sample of student responses to share with a colleague.

Consider using this data-gathering activity at the beginning of the month and repeating it at the end of the month to look for growth in students' beliefs about mistakes.

OPTION 3: CLASSROOM DISCUSSION OR INTERVIEW.

"Mistakes: Are they good? Bad? What do they have to do with learning?"

Start your math class with a brief discussion of the questions above or interview a few students one on one to get a sampling of what students are thinking.

Record responses from students to reflect on with a colleague at next month's meeting.

ADDITIONAL PROFESSIONAL LEARNING EXPERIENCES

The additional professional learning experiences for November relate to addition and subtraction. An open-number-line model is used to help students develop flexible computational strategies based on place-value understandings and number sense. As students learn to add and subtract using algorithms, often they make calculation errors because they are using an abstract procedure they have yet to internalize. When students use an open number line as a tool to verify or explain their answers, they can catch and correct their mistakes.

Math Talks: Number Talk

This month's math talk is a number talk, a five- to 15-minute classroom routine in which students are presented with a computational problem to solve mentally followed by a discussion of solution strategies.

Here are some other important features of number talks:

- The teacher is a facilitator and recorder (refrain from teaching or determining correctness)

- All solutions that are shared are recorded and students are asked to defend their answers (this is an opportunity for students to determine correctness and find and fix their mistakes)

To watch a number talk in action, visit **youcubed.org. https://bit.ly/4c7xYfa**

Together as a math teaching community, give this routine a try. You can use a computational problem similar to problems you are currently using with students, or you might use this problem:

$$91 - 57 =$$

Discuss the experience:

- What different strategies did we use to solve the problem?

- How might a focus on solving the problem in different ways allow others to find and correct their mistakes and to see mistakes as learning opportunities?

- How might you use a number talk in your classroom?

Manipulatives and Models Matter: Open Number Lines

If we believe mistakes are an essential part of the learning process, we can help students learn how to recognize and correct their own mistakes. Manipulatives and models offer our students the opportunity to see math in action. As students move between concrete experiences, representations, and abstract ideas, they check their thinking against the manipulatives and models as they prove or disprove their solutions.

Take subtraction, for example. Students write equations to represent subtraction computations. An open number line can help students verify their solutions. Imagine that your student has solved a problem and came up with the following answer:

100 – 53 = 57

Your student reasons that 50 + 50 makes 100 and that 3 + 7 makes 10. They use the relationship between addition and subtraction, their understanding of base ten, and their partners to ten to support their thinking. Now, you ask the student to verify their solution on the open number line.

They start at 53 and talk about how they will jump ahead to 100. Notice the jumps that are made below.

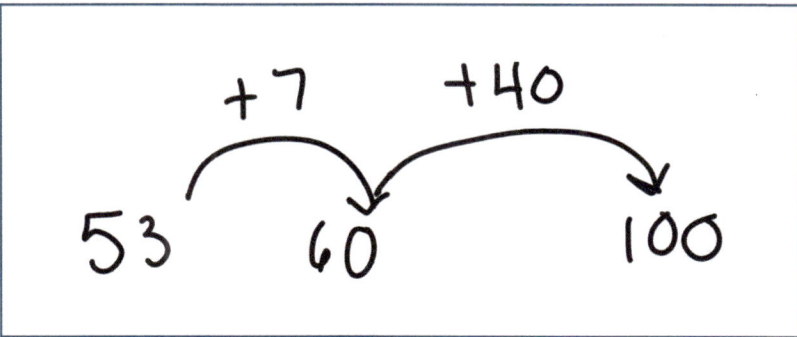

The student uses partners to ten (3 + 7) as they can jump to the next ten. Then, they use similar thinking to jump from 60 to 100 (6 + 4 = 10 so 60 + 40 = 100). Recognizing that the jump from 60 to 100 is 40, the student now sees that the difference is 47 (7 + 40) instead of 57. The open number line allows the student to find and correct their mistake.

(Continued)

(Continued)

You might follow up by asking, "How did using the open number line help to change your thinking?"

Try this experience with your math teaching community.

Select two problems below and practice adding or subtracting using an open number line. Share your results with a partner.

48 + 19 =

71 – 33 =

148 + 39 =

382 – 27 =

1,259 + 132 =

4,510 – 619 =

REFLECT

- How did the experience of using an open number line take you through the learning cycle (understand, try it out, analyze, try again)?

- How might use of the open number line and other mathematical models and tools help students to see mistakes as a natural part of learning?

Game Time

This month's game links the experiences from the Math Talk and the Manipulatives and Models Matter sections by extending the work with open number lines with subtraction.

ZERO!

Materials:

2 players

Dice (1 per player)

Pencil

Zero! recording sheet

Zero! Version 2 recording sheet

Zero!

ROLL	VALUE	OPEN NUMBER LINE	EQUATION
			100- _____ =_____

Challenge: Be the player to get closest to zero without going under.

Game directions:

1. On your turn, roll the die and record the result as a ones or tens.

2. Subtract the selected number from 100 using an open number line.

3. Write an equation, and the difference is now your new starting number.

4. It's now your opponent's turn. They will repeat steps 1–3.

5. Both you and your opponent must use 6 rolls.

6. The winner is the player closest to zero after 6 turns.

Version 2:

In this version of the game, roll two dice and create a two-digit number. Subtract your number from 100 using an open number line. Game play continues back and forth until you and your opponent have rolled six times. The player closest to zero without going under is the winner.

(Continued)

(Continued)

Zero! Version 2

Roll 2 dice. Use one digit as ones and the other digit as tens.

ROLL	NUMBER	OPEN NUMBER LINE	EQUATION
_____ , _____			100 – _____ = _____
_____ , _____			
_____ , _____			
_____ , _____			
_____ , _____			
_____ , _____			

Total _____

Player 1 is_____away from zero. Player 2 is_____away from zero.

Discuss the experience:

- How does this game reinforce subtraction strategies?

- What math habits did you use while playing the game?

- How might you adapt this game for use with your students?

- How can games like *Zero!* help students see mistakes as a natural part of the learning process? How can they help students develop the habit of verifying their thinking to catch and correct their mistakes?

 Available as a downloadable resource on the companion website.

IN YOUR CLASSROOM

Classroom Story

Mrs. Jones's class of third graders is a wonderfully diverse group of chatty eight- and nine-year-olds. All year, they have had opportunities to share their thinking in math through number talks, daily number sense routines, and open tasks in which students grapple with ideas and make sense of math. One morning, students were engaging in fraction tasks to build a deeper understanding of the relationship between the numerator and the denominator. Before they started their lesson, Mrs. Jones wanted to determine how students were currently thinking about fractional relationships. She asked the students to consider the square below and think about how much of the whole square was shaded.

How much of the whole is shaded? How do you know?

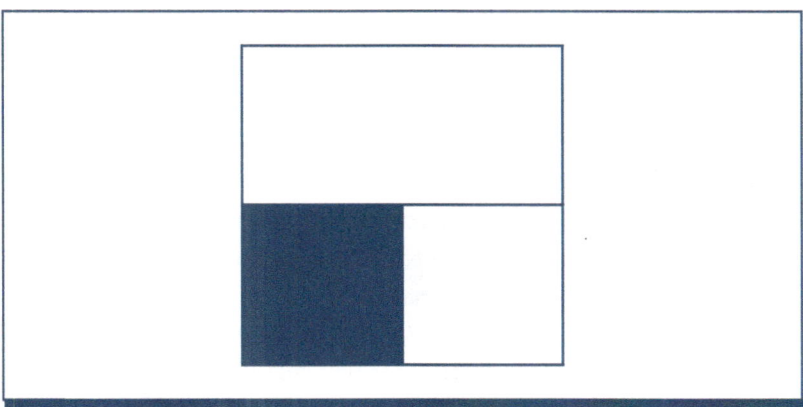

After students had a minute to think, Mrs. Jones asked them to turn and talk to their neighbor about their response and reasoning. Students excitedly turned to share their thoughts. Mrs. Jones listened in to gain a sense of the students' current understandings and make some decisions about how she would stage the class discussion. Once students had a minute to share, they were redirected back to the board for a whole-group share.

Mrs. Jones asked if anyone had an answer they would like written on the board that described the shaded area. She called on Sam first, knowing that he had one of the more common answers during the pair–share.

Sam:	I think it's one third. There are three sections, and one of them is shaded. A few students around the room nodded their heads in agreement. Mrs. Jones wrote Sam's answer on the board next to the square.
Mrs. Jones:	It seems like there are a number of people who also said one third. Did anyone else have a different answer that names the shaded part of the whole? Many students raised their hands. Mila shared next.
Mila:	I think that the shaded part is one fourth. If you imagine there is a line going up in the middle, you can see the four parts.
Mrs. Jones:	I see many of you agreeing with Mila. Would someone like to add on to what Mila said?
Cora:	I also think it's one fourth.
Mrs. Jones:	Why do you think that?
Cora:	The two bottom squares look the same. It would be the same on top. So there are four, and one of them is shaded in.

Mrs. Jones wrote one fourth next to the one third. She then asked the class if there were any other answers. No other answers were given, so Mrs. Jones said, "So it looks like we agree that it is either one third or one fourth. Can it be both?"

The students all agreed that it had to be one or the other. A number of other hands went up as students tried to justify why either one third or one fourth named the shaded part. Many students changed their minds, becoming convinced that the shaded part represented one fourth of the whole. Sam, however, was holding strong.

Sam: *I still think it's one third. We learned that the bottom number is the number of parts and there are three sections. So, it should be one third.*

Mrs. Jones understood that students still needed some exploration regarding representing fractions as a relationship between the parts and the whole. They would be working with Cuisenaire rods that day to think about parts of a whole and how we represent a part of a whole with a fractional name.

When we set the expectation that all students will share their math thinking, our students see that everyone's ideas contribute to the class's growing mathematical understanding. When we accept incorrect responses along with correct ones, students learn that mistakes are a natural part of learning. If our goal is to build a school where students enjoy engaging in mathematics and flourish, we must consciously and intentionally listen to our student's thinking.

Anchor Lesson

Day 1: Naming Feelings Around Being Incorrect

WHAT?

This lesson will help students begin to shift their mindset about how we think about and respond to mistakes in math.

YOU NEED:

Challenging math problem with incorrect solution(s)

> Chart paper
>
> Markers

HOW?

1. Choose a problem related to your class's current math content focus. Create an incorrect solution to the problem. You could use a common misconception or computational error.

2. Show the problem and incorrect answer to the class, and facilitate a discussion around the correct answer and why a student might have made this mistake. Ask students to imagine how a student who made this mistake might feel. Have students turn and talk about how they felt when they made mistakes in the past. Talk with students about how these feelings (e.g., embarrassment, sadness, anxiety, disappointment, frustration, confusion) impact learning. Explain that mistakes are a good thing because they help us learn. When a class member makes a mistake, it helps us all to use our math brains to become stronger mathematicians. (See the Spotlight on Brain Science in this chapter, pp. 166–167.)

3. Elicit ideas from students about what they can say and how they can respond when someone makes a mistake. Record these ideas on an anchor chart titled "I can support others when we make a mistake." Introduce the following sentence frames, and invite students to think about how the class can use statements like these to support themselves and others as learners. Challenge students to try out these sentence frames during class discussions, while journaling, or when they reflect on their learning. Encourage students to think of additional sentence frames that can help the class think about mistakes as a natural and needed part of learning.

> At first, I thought _____, but now I think _____.
>
> This mistake is helping me to think about _____.
>
> I'm learning _____.

REFLECT

Many people believe that mistakes should be avoided in math and that if they make a mistake, it means they aren't good at math. This chapter has highlighted the importance of making and recognizing mistakes to deeply learning mathematics. We can shift our own mindsets and learn to accept and celebrate mistakes in the classroom. In turn, our students can see their mistakes as a powerful tool for learning.

Use the discussion questions to reflect with students:

- What might be challenging for you when you make a mistake?

- How will supporting each other through mistakes help our math community?

Day 2: How Do We Disagree Respectfully?

WHAT?

In part 2 of this lesson, students will think about actionable steps they can take to disagree respectfully with their peers.

YOU NEED:

Chart paper

Markers

HOW?

1. Write "How do we disagree respectfully?" at the top of an anchor chart.

2. Remind students that doing math in the classroom means that they will share their ideas and solutions out loud and in writing. You might say, "There will be times when we disagree on an answer. This is part of learning and can help us all to understand math ideas. It's important to think about how we will disagree respectfully to support each other as learners."

3. Ask students to share phrases they can use when they disagree with another student's idea.

   a. Some common responses might include:

      • Critique the idea, not the person.

      • Use language such as "I disagree with that idea."

      • Point out the parts that are correct.

      • Thank your peers for helping you learn.

4. Consider providing some sentence stems that students can use when disagreeing with an answer or explanation.

   • I disagree with _____'s idea because _____.

   • I had a different solution from _____'s solution.

   • I can see part of _____'s solution is correct, but I think _____.

5. Display the chart for students to refer to. Make sure to practice the phrases and provide feedback to students.

REFLECT

When our students begin to recognize and celebrate mistakes, they will need to practice responding graciously. We can help our students notice and name when they disagree and brainstorm responses that maintain the integrity of the classroom community.

Use the discussion questions to reflect with students:

• Why might it be an important skill to disagree with someone in a respectful way?

• How will using the sentence stems we created support our classroom math community?

LET'S DO SOME MATH ACROSS OUR SCHOOL!

We feel less concerned with mistakes when we experience the joy of creating something new and beautiful. In the creation of a quilt, we can bring our own artistic flair into the design while using mathematics to pattern, find symmetry, and calculate sizing. Just as you and your colleagues did earlier, have your students create a math quilt that applies their mathematical understanding and gives them the freedom of design where making a mistake doesn't instill fear.

Tell students that they will engage in a creative activity in which they can use their math abilities and that you want them to pay attention to their feelings as they work on this mathematical problem. Share the sample quilt with students and discuss the mathematics they see. Select one of the quilt problems below for students to solve. Once they have had the opportunity to create their designs or solve their problem, you might have students share with their peers. Invite them to leave a comment on what was most appealing to them.

GRADES K-1

Create a pattern in the quilt and share how you know it's a pattern.

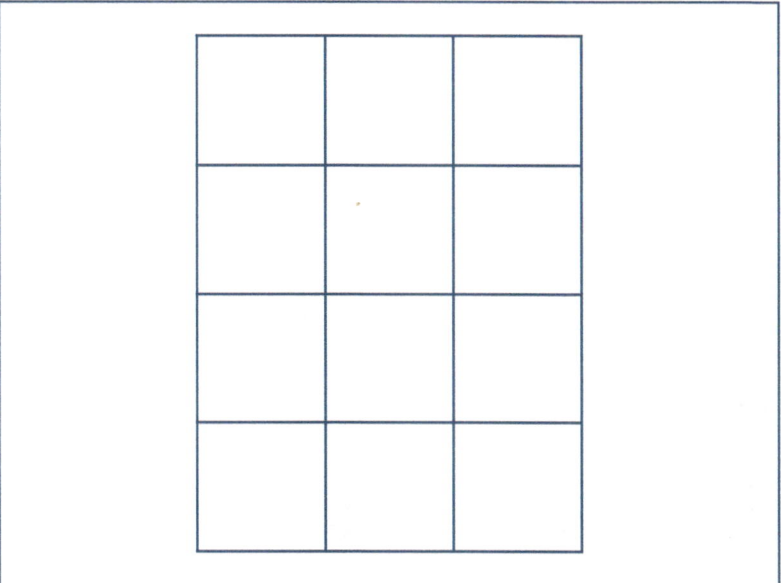

(Continued)

(Continued)

GRADES 2-3

Option 1: If I square is 4 inches long, how long is the quilt from left to right? Top to bottom? Share how you know.

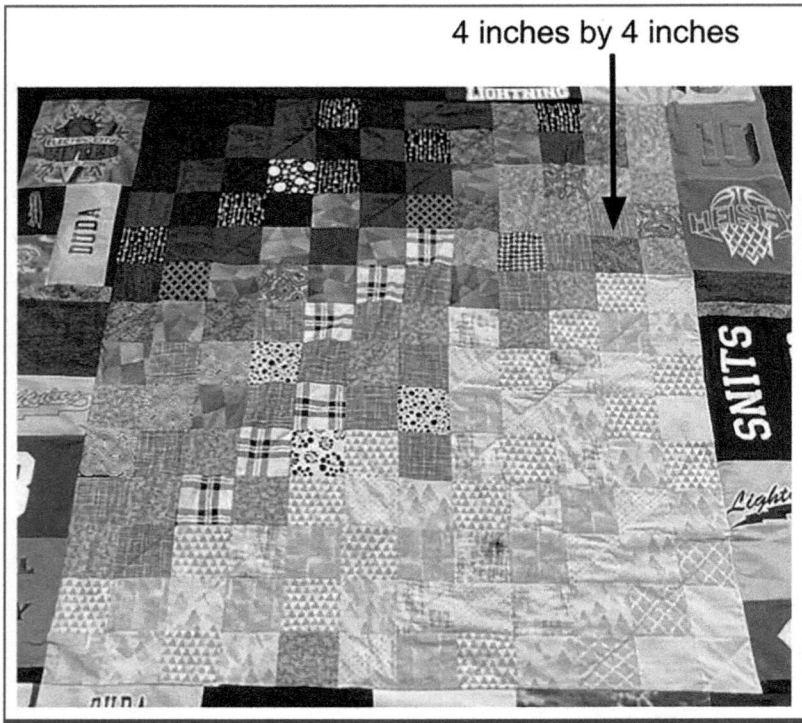

4 inches by 4 inches

Option 2: Create a pattern in which half the interior squares are one color and half are another color. Share how you know each color represents half of the quilt.

GRADES 4-5

A sample quilt design is shown in Figure 5.2. If each square in the border should be the same, what are the dimensions that each t-shirt should be cut to? Include a half-inch for sewing. (See Figure 5.3 for a completed design.)

Figure 5.2 ◆ Sample Design

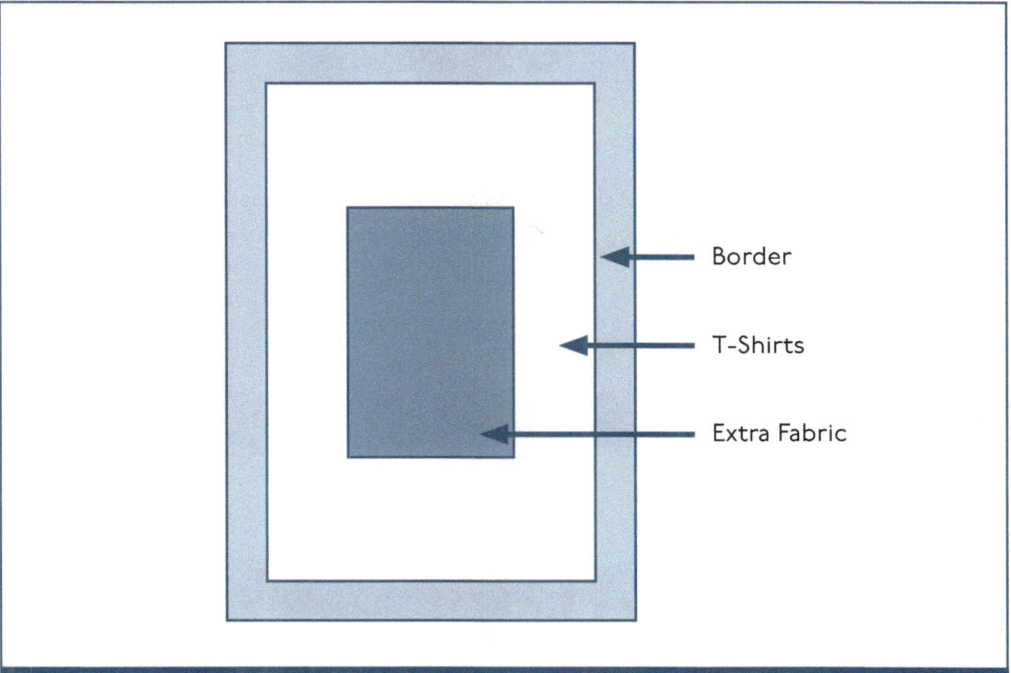

Border

T-Shirts

Extra Fabric

Figure 5.3 ◆ Completed Design

(Continued)

(Continued)

Reflection Questions:

- How did you use math as you designed your quilt?

- What are some feelings you experienced as you created your quilt? Were you worried about making mistakes? Why or why not?

- What are some things we can do as a class so that our math learning work feels creative and satisfying like we're designing a quilt?

 Available as a downloadable resource on the companion website.

TEACHING MOVE: RECORD ALL RESPONSES WITHOUT JUDGMENT

Students are fantastic at reading body language and cues. They learn early on that if they provide an answer and the teacher doesn't record it on the board or hesitates and stops recording their ideas, it must mean they are incorrect. We can help students see mistakes as opportunities for learning by consciously refraining from providing these cues and instead recording all answers on the board for consideration and discussion by the class. In the classroom vignette earlier, Mrs. Jones exemplified the practice as she wrote correct and incorrect student responses on the board without judgment. When she recorded Sam's incorrect response, Mrs. Jones recognized and positioned Sam as a mathematical thinker. She allowed him to analyze the correctness of his and other students' responses and, in doing so, communicated her belief in Sam's ability to understand mathematical ideas.

REFLECT

- What might be the impact on student learning if the teacher records all responses without judgment?

TEACHING MOVE: DEFER TO THE CLASS TO DETERMINE THE CORRECTNESS OF AN ANSWER

Another teacher practice that uses mistakes as opportunities to learn is to defer back to the class to determine the correctness of an answer. Think back to Mrs. Jones's class in which students were determining the shaded part of the whole.

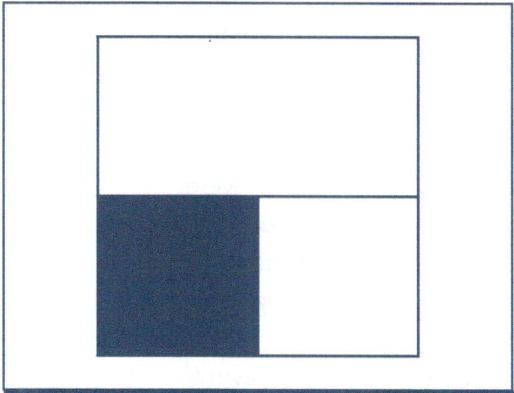

Mrs. Jones intentionally placed the burden of finding and correcting mistakes back on to her students. She knew that students would justify their answers. To do so, Mrs. Jones asked the question "Can both answers be correct?" By asking this question, students naturally began defending their choices, leading students who were incorrect to correct solutions and understand why the correct answer was correct.

It's important to highlight the power of opening up discussion to the class, especially when there is a difference in answers. Students have an incredible ability to reason through their ideas. Any route they may have taken to arrive at an answer, whether correct or incorrect, is valid in forming a deep understanding of the mathematics at hand.

REFLECT

- How might students feel differently about their mistakes if we open answers up to the class for discussion?

TEACHING MOVE: PUSH FOR DEEPER UNDERSTANDING WITH BOTH RIGHT AND WRONG ANSWERS

Imagine a teacher walking past a student's desk. They pause and say, "How did you get that answer?" What happens next? It's not hard to imagine the student erasing their answer, assuming it must be wrong.

Students are constantly seeking feedback, but our feedback needs to be about more than just the correctness of answers. Feedback can help students reflect on and justify their answers and refine their work.

When a student brings us work, correct or incorrect, we can pause and ask a question. We might say, "Tell me about what you did." Several things might happen. As the student describes their work, they may recognize a part where what they thought and what they wrote aren't the same. Or they may realize that their ideas aren't finished and they need to go back and think again. We can prompt again with questions such as "What might you try now?" or "What will you add to your work?" Students may also feel affirmed in their answers as they restate what they did and how they did it. In each case, the teacher is listening and learning from students about their thought process, their understanding, and misunderstandings.

REFLECT

- How might pushing for deeper thinking with both right and wrong answers support student learning?

CLASSROOM ROUTINE: MY FAVORITE NO

My Favorite No helps students practice thinking about mistakes as stepping stones to learning.

Use these steps to get your class started with My Favorite No.

1. Pose a question or problem.

2. Have students record their responses on notecards.

3. After the lesson, sort the cards into two piles: correct and incorrect, and select an incorrect response to be used as "my favorite no" for the next day.

During the next lesson, rewrite the problem with the incorrect response to be shared using a document camera, chart paper, etc. Ask, "What is done correctly?" and invite the class to discuss the strengths of the work. Then ask, "What is incorrect? How do you know?" and encourage students to collaborate in respectfully critiquing and improving the work.

Note that students' responses to the initial question or problem also provide important formative assessment data that can be used to determine small groups and next steps in instruction.

STATION ACTIVITY: WHICH ONE IS IT NOT?

We can help students see mistakes as helpful for learning when we ask them to analyze incorrect solutions to problems. For this month, the station activity focuses students on looking at and thinking about nonexamples through incorrect answers.

Using the cards provided (available online) or others that you create, students select a card with a math problem and multiple-choice answers. They choose an incorrect answer and use the "Which One Is It Not?" recording sheet to justify why that answer choice is incorrect.

Example:

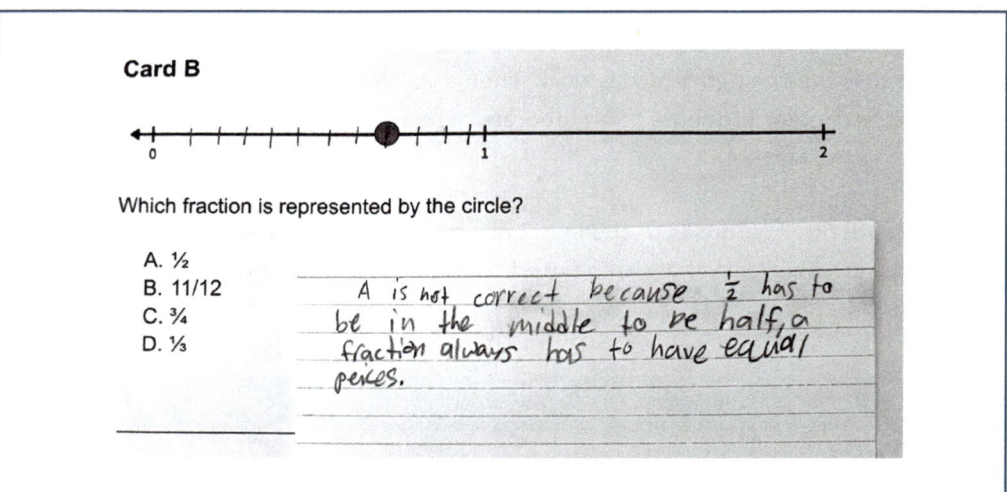

WHICH ONE IS IT NOT?

Select a card from the stack of cards. Read the question and answer choices. Select ONE answer that is NOT correct and complete the sentence stem.

_____ is not correct because _____.

Share your response with someone who answered the same card as you. Compare your reasoning.

As students engage in this station activity, they consider how incorrect answers are related to correct answers.

Station activity responses can also be used for class discussions and small-group instruction.

 Available as a downloadable resource on the companion website.

LITERATURE CONNECTIONS

It's Okay to Make Mistakes **by Parr** (2014) (Grades K–2)

This colorful book shares everyday examples of common mistakes and assures readers/listeners that these mistakes are okay. The author's parting message is this: "It's okay to make mistakes sometimes. Everyone does—even grown-ups! That's how we learn."

Activity: What is a mistake that you have made? How did it help you learn?

Next time you or someone else makes a mistake, say, "It's okay to make a mistake. That's how we learn."

A Whale of a Mistake **by Hobai** (2020) (Grades 3–5)

Through a series of metaphors and beautiful illustrations, this book helps readers/listeners reflect on the feelings that mistakes can elicit and consider ways to address these emotions so that mistakes are seen as a natural part of life.

Activity: What's a mistake you made in the past that helped you to learn?

When you make a mistake in math or another part of your life, what are some ways you can "wave" to that mistake and then move on?

How are you showing courage when you face and learn from mistakes?

SPOTLIGHT ON BRAIN RESEARCH: MISTAKES HELP US LEARN

If students believe that they can get better at math and their effort is worthwhile, they come to see mistakes as a welcome occurrence. Boaler (2022) wrote a lot about a growth mindset in mathematics in her book *Mathematical Mindsets*. She shared that our brains "spark and grow" when we make mistakes and that teachers should use these opportunities to the student's advantage (p. 13).

You can share the following information with students in a mini lesson or a station activity. You might also send this reproducible home along with the Family Newsletter to help families talk about these important ideas.

MISTAKES HELP US LEARN.

These photos show an electrical brain current called a "synapse." Synapses fire back and forth in our brains all the time when we are playing and working, creating connections among the neurons of our brain. When a synapse fires in our brains, we are building connections in our brain pathway and learning. These connections between neurons are always changing throughout our lives.

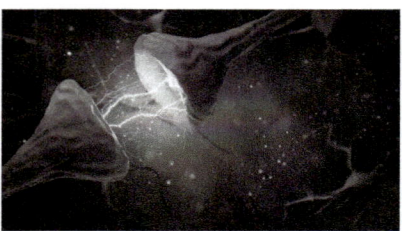

Image source: Istock.com/K_E_N

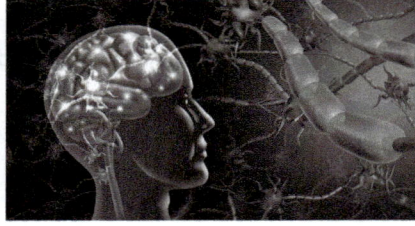

Image source: Istock.com/wildpixel

Our brain fires a synapse when it is trying to decide if something makes sense. It fires another synapse when it sees that we have made a mistake. So, making a mistake is a very good thing for doing math because it helps us to learn.

Choose one of these ways to think more about this important information:

1. Write down why these ideas are important to you and how you want to put them to work in your life.

2. Talk to a friend about these ideas. Decide together how you will put them to work.

3. Share these ideas with someone at home. Tell them why these ideas are important to you and how you plan to use them.

 Available as a downloadable resource on the companion website.

SPOTLIGHT ON EQUITY: SAFE TO MAKE MISTAKES

If we want students to view mistakes as stepping stones to learning, it is essential that they feel safe making mistakes and discussing their mistakes with others. Hammond (2014) explained that the feeling of psychological safety originates in our brain. When we feel socially, emotionally, or intellectually unsafe, our brains produce hormones that shut down critical thinking and make learning difficult. However, when we feel connected to others, our brain triggers the release of different hormones that support learning. A caring and connected learning community is essential to academic risk taking and acceptance of mistakes as a natural part of the learning process.

TRY IT

Here are some ways we can learn more about our students' feelings of social, emotional, and intellectual safety in our math classroom:

- Conduct a survey.
  - On a scale of 0–10, how safe do you feel to share your math thinking in our classroom? Why do you feel this way?
  - On a scale of 0–10, how safe do you think other students feel sharing their math thinking? Why do you feel this way?
- Start a class discussion.
  - What would help you to feel safe in math class?
  - What might help you or others to feel comfortable enough to raise your hand and share your math thinking?
  - When you share your math thinking during a class discussion, how do you know that the answer you give is accepted and respected, even if it is wrong?

Create a class anchor chart titled "Creating a safe space in our math classroom." Gather students' ideas and record them on the chart for all students to reference.

MATHEMATICAL ME: STUDENT JOURNAL AND PORTFOLIO

Student Journals:

Have students respond to the following questions:

• How do mistakes help me learn math?

• What should I say or do when I make a math mistake?

• What should I say or do when someone else makes a math mistake?

Older students can write their responses in their Mathematical Me journals. Younger students can draw a picture showing their response, or you might gather responses during a class discussion or quick one-on-one interviews.

Review students' responses to monitor students' beliefs about the role of mistakes in learning. You might summarize this data by tallying or graphing the different ideas students offer.

Be sure to share and discuss this data with your class. Ask, "What does this data tell us?" and "What are some things we can do to continue growing as mathematicians?"

Student Portfolios:

Have students choose a piece of work where they made and corrected a mistake. On a sticky note, have them write the date and complete this sentence frame:

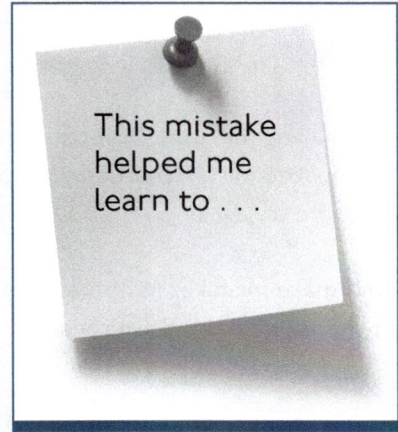

This mistake helped me learn to . . .

Source: Istock.com/Thammask-Chuenchom

FAMILY NEWSLETTER: WE LEARN FROM MATH MISTAKES!

WHAT WE'RE LEARNING

This month we're learning about how the mistakes we make while doing mathematics help us to learn math at a deep level.

WHY IT'S IMPORTANT

The word "mistake" carries a negative connotation. Feelings of shame and embarrassment often surface when errors are made, especially in mathematics. These negative feelings make it difficult to learn. Your child needs to understand that mistakes are essential to their learning.

Brain research has shown that mistakes actually benefit us. When we make a mistake, our brains fire electrical impulses or synapses that heighten our awareness and readiness for learning.

When you help your child to see that mistakes support learning, you set them up for future success in school and in life.

HOW YOU CAN HELP

Engage your child in a conversation about making mistakes. Ask your child how they feel when they make a mistake. Remind them that we all make mistakes on a regular basis and mistakes help us learn and grow.

Allow your child to see you making mistakes. Use clear and positive language to describe the mistake you made and how the mistake helped you to grow. You might say, "Whoops. I didn't do that correctly. Let me try again."

Knowing that your child will make mistakes at home, plan for how you will respond when mistakes are made. You might ask questions like these:

- "What did you learn?"

- "What will you try differently?"

 Available as a downloadable resource on the companion website.

CHECKING IN ON OUR LEARNING: WE LEARN FROM MATH MISTAKES!

- How do mistakes help me learn math?

- What should I say or do when I make a math mistake?

- What should I say or do when someone else makes a math mistake?

This month's learning focus, making mistakes, is pivotal in building a powerful math community. Our brains use mistakes to create strong neural pathways, allowing us to learn and understand mathematics at a much deeper level. In helping students see mistakes as valuable, teachers create safe spaces that combat shame and negative feelings regarding a natural part of learning.

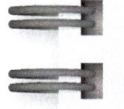

MATHEMATICAL ME: EDUCATOR JOURNAL

- How has this month's learning focus supported your students' mathematical growth?

- How has this month's learning focus supported your growth as a math teacher?

- How has this month's learning focus supported your school's growth as a powerful math community?

LOOKING AHEAD

In December, we will explore how our mathematical identities support math learning and how educators can grow students' math identities.

Questions for Reflection

- Do you consider yourself a mathematician? Why or why not?

- Why is this an important question for educators to consider?

CHAPTER SIX

December
We See Ourselves as Mathematicians!

ESSENTIAL QUESTIONS

- What does it mean to be a mathematician?
- How do I know I am a mathematician?

THIS MONTH'S FOCUS

In a powerful math community, students and educators see each other as mathematicians, as people who engage in mathematical activity in and out of school because mathematics is useful, interesting, and fun.

Educators take responsibility for helping all students develop their math identities and agency in addition to mathematical understandings and proficiencies.

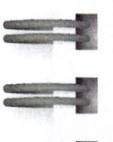

MATHEMATICAL ME: EDUCATOR JOURNAL

Take five minutes to respond to the essential questions. You will have a chance to reflect on these same questions and to have students respond to these questions at the end of the month.

ON YOUR OWN

LET'S DO SOME MATH!
A TRUE MATH STORY ABOUT ME

Write a true math story problem to help others learn more about you and how you use mathematics outside of school. Consider how you engaged in the habits of mathematically powerful people as you wrote and solved your true math story problem.

 online resources   Available as a downloadable resource on the companion website.

Sue's example: I enjoy working with clay and want to begin a new clay project. I've been thinking about making a clay garden totem for my backyard, but I want to get a rough idea of how much it will cost before starting this big venture. Clay costs $17.25 for a 25-pound slab. Each totem piece will likely be about six inches tall and require close to one pound of clay. Based on my past clay projects, I expect that firing and glazing each piece will cost about $2.00. I want a totem that is approximately three feet high. From my latest visit to the hardware store, I know the rod and bolts needed to assemble the totem will cost about $20.00. What's a reasonable estimate for the final cost of the garden totem?

Why This Focus

Sue's Math Story:

We hear a lot about how mathematical proficiency opens doors for students to future math-related academic and career opportunities. Although I believe this cause–effect relationship is important, I think it misses a critical piece of the big picture about why math is important. My son and husband might well have chosen careers in engineering if they had experienced better math instruction in elementary and secondary school. They both have personal passions that involve mathematics that they now pursue as hobbies outside of work—repairing a 30-year-old truck, building a piece of furniture to meet a unique need, designing a gardening system, perfecting a recipe for barbecue ribs. In addition, they each found a way to satisfy their interest in math within their career, my husband as a financial analyst for NASA and my son as a research psychologist. On the other hand, my daughter, I believe, was always destined to become a teacher. The world is without question a better place because she chose to teach high school English as a way of helping students learn about themselves and the world. For my daughter, mathematics is a tool rather than a passion. Mathematics helps her daily to be more effective as a teacher and to navigate life strategically.

The reason we need to provide every student with high-quality mathematics learning experiences, pre-K through 12th grade, goes well beyond preparing the next generation for STEM careers, and it certainly goes beyond helping students to pass achievement tests. Learning to use mathematics as a lens for understanding and interacting with the world will enrich our students' lives, helping them to see and do things they wouldn't be able to see and do otherwise. When our students recognize themselves as mathematical beings, they are better positioned to flourish as individuals and to contribute to their communities and our world.

And so, we must begin our work in support of our students' math identities by looking in the mirror. We can't effectively position our students as mathematicians unless we first recognize ourselves as mathematicians. Helping our students to grow strong math identities begins with our own math identities.

Tapping Into Your Experience

When we consciously don the identity of a mathematician, we notice how mathematics supports us moment to moment and how we are richer because of our mathematical lives.

Before meeting with your team or faculty, use this exercise to look at your daily life through a mathematical lens to articulate and strengthen your own mathematical identity.

> We must begin our work in support of our students' math identities by looking in the mirror. We can't effectively position our students as mathematicians unless we first recognize ourselves as mathematicians. Helping our students to grow strong math identities begins with our own math identities.

The habits of mathematically powerful people (Figure 6.1) can be thought of as a description of what it means to be a mathematician. What are some ways you engage in these habits as a part of your daily life? What specific examples can you identify for each habit? Post your list in a visible location so you can add more examples as you notice them.

Figure 6.1 • Habits of Mathematically Powerful People

I expect math to make sense.

I love challenging math problems.

I learn from math mistakes.

I see myself as a mathematician.

I talk about math.

I represent math in different ways.

I make math connections.

I look for and use patterns.

- I expect math to make sense.

  Example: When my monthly water bill is higher than expected, I consider and investigate possible causes.

- I love challenging math problems.

  Example: I create a personal budget that takes into account multiple family members' needs, the realities of inflation, and the goal of financial security.

- I learn from math mistakes.

  Example: When I make an error in estimating the amount of time it will take to drive across town, I use what I learn to make more accurate calculations in the future.

- I see myself as a mathematician.

  Example: I research and analyze data about the benefits of different types of exercise and recommended workout times to plan a daily fitness routine.

- I talk about math.

  Example: I talk with my insurance agent and family members about options and costs for different insurance plans.

- I represent math in different ways.

  Example: As I plan my spring garden, I place stakes in the ground to determine its dimensions, sketch a diagram showing how the vegetables will be arranged, and then calculate the quantity and cost of the soil and plants.

- I make math connections.

  Example: I use my experience calculating tips to think through the amount of interest I will pay on a car loan.

- I look for and use patterns.

  Example: I use temperature and rainfall patterns to decide on the best time of year to take a special vacation.

Mathematical Identity

In the National Council of Teachers of Mathematics (NCTM) publication *The Impact of Identity in K-8 Mathematics*, Aguirre et al. (2013) defined mathematics identity as "the dispositions and deeply held beliefs that students develop about their ability to participate and perform effectively in mathematical contexts and to use mathematics in powerful ways across the context of their lives" (p. 14) or, more simply, "the stories that people tell about themselves and what they view as important" (p. 27).

- How might you define the concept of mathematical identity in your own words to explain this concept to a colleague, a parent, or students?

- Why is the concept of mathematical identity important for educators, parents, and students to think about and discuss?

Mathematics as a Habit

In his book *Atomic Habits*, James Clear defined a habit as "a routine or practice performed regularly; an automatic response to a specific situation" (2018, n.p.).

- Using this definition, how might you think about using or doing mathematics as a habit? What does a math habit look and sound like for a medical professional? A small business owner? A parent? A teacher?

- If educators considered mathematics as a habit in the same way that reading is often considered a habit, how might this change how we approach mathematics instruction?

When we view mathematics as a habit that can be learned and strengthened through practice, we position all students as mathematically capable and destined to continue growing their mathematical capabilities through practice. Clear observed, "Building habits in the present allows you to do more of what you want in the future" (p. 47). When mathematics is understood as a way of looking at and interacting with the world, we habitually employ mathematical ways of thinking and acting that allow us to tap into our best critical thinking. And as we regularly engage in the habits of mathematically powerful people, we simultaneously reinforce our mathematical identities.

Identities evolve over time based on the meaning we make of our experiences. We can grow identities that align with our core values and help students grow their desired identities by attending to our daily habits. According to habit scientist Clear, "Your identity emerges out of your habits" (p. 41). In addition, our identities motivate us to further develop and refine our habits (see Figure 6.2).

Figure 6.2 • Our Mathematical Habits and Identity

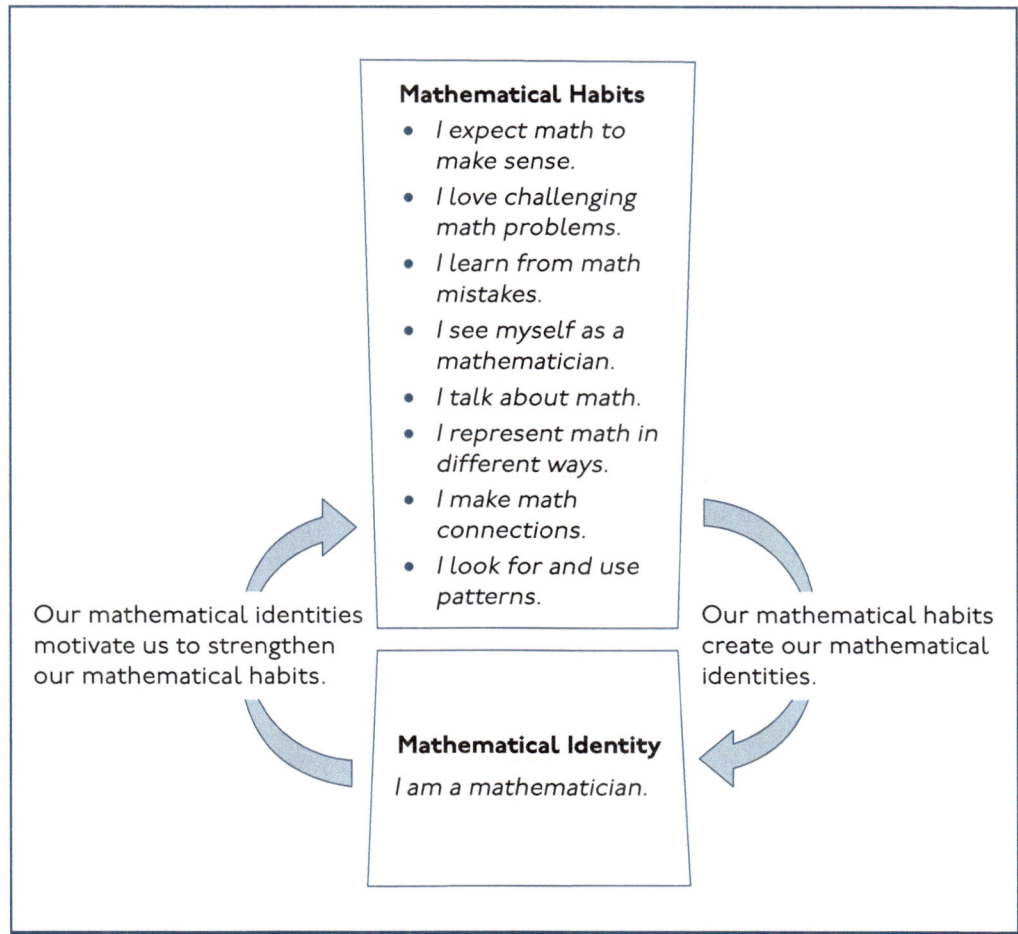

As we develop our math habits, we come to see ourselves as mathematicians. These mathematical identities, in turn, lead us to take actions that strengthen our math habits. Individual teachers, schools, and school districts—we need to all hold ourselves accountable for cultivating math as a habit and an identity in every one of our students. We achieve this important goal through the following three essential steps:

1. Intentionally design and implement learning experiences that allow students to grow the habits of mathematically powerful people.

2. Provide structured opportunities for students to learn about and reflect on their mathematical identities.

3. Set students up for future learning by teaching them how to practice and refine their math habits in order to grow their desired mathematical identities.

As we develop our math habits, we come to see ourselves as mathematicians. These mathematical identities, in turn, lead us to take actions that strengthen our math habits.

Our job as educators is not complete until we have achieved Step 3. The process of growing our math habits and strengthening our mathematical identities is continuous and ever evolving. We need to equip our students to continue their learning journey long after they've left our classrooms.

Culturally Relevant Math Tasks

In their book *Engaging in Culturally Relevant Math Tasks*, Matthews et al. (2022) stated, "Children often learn mathematics is not real—and when it is, it has nothing to do with them" (p. 40). If we want our students to see themselves as mathematicians, as people who use mathematics in and out of school, we need to regularly provide them with mathematics tasks that are relevant to their lives. Matthews et al. encouraged us to use the following guiding questions in choosing and designing the mathematics tasks to use with students:

> Demand—How does the task focus on building deep conceptual knowledge and prompt children to do and create mathematical knowledge?
>
> Relevance—How is the context and mathematical inquiry rooted in affirming and exploring cultural knowledges and identities? Does the task context and prompt feature empowered relationships, understandings about their community and themselves?
>
> Agency—How does the task prompt and context push students to respond to needs and issues with empathy, critical consciousness and social action? (2022, p. 35)

Consider how the following mathematics problem from Matthews et al. might meet these criteria:

> Malik and Joel have created a volunteer project to mow the lawns of senior citizens in their neighborhood. In the first hour Malik mowed 5 of 8 rows of the lawn at one house, and Joel mowed 3 of 7 rows of the lawn at another house. How much of the two lawns have they mowed so far? What fraction of the two lawns still need to be mowed? How will Malik and Joel's partnership support the neighborhood? (2022, p. 38)

Identify a mathematics task you will be using in the next week. How might you strengthen the task by using the criteria of demand, relevance, and agency to help your students see themselves in this mathematics task?

TOGETHER WITH YOUR TEACHING COMMUNITY

Since We Met Last (10 minutes)

In November, you and your colleagues collected some classroom data about how students think about mistakes in the context of their learning.

Use the following protocol to share and reflect on this data with your math teaching community:

- Share your data with a colleague. (5 minutes)

- As a group, discuss what you learned from the data and how these insights can help you to support your students' growth as mathematicians. (5 minutes)

Let's Do Some Math Together! (10 minutes)

Trade your true math story problem with a colleague. Think about and solve each other's problems. Share your solutions with each other.

Discuss:

1. What did you learn about yourself and your colleague from this experience?

2. How did this experience help you to think about what it means to be a person who consciously uses math in real life? Why is this important?

Building Our Expertise (30 minutes)

Research psychologist Fredrickson (2009) has identified 10 positive emotions that, when experienced regularly, help people to thrive in life: joy, pride, gratitude, amusement, serenity, awe, interest, inspiration, hope, and love. Choose one of these emotions

(Continued)

(Continued)

that you have experienced as a math teacher during the first half of the school year. Share with a colleague. Reflect together about how your own math identity has grown stronger this semester.

Our math identity and our students' math identities are intertwined. Our students' math identities are ever evolving as a result of their experiences with mathematics and their interactions with fellow mathematicians, classmates, teachers, and family members.

Our students' math identities are an important focus for efforts to create equitable and inclusive math classrooms. Therefore, we need to monitor and celebrate our students' and our own growing math identities.

Shirley and Hargreaves observed that teachers play an important role in building their students' identities. They reminded us that:

> Educators know all too well the phenomenon of young people being or becoming far more than they have always seemed. The struggling student who becomes a late bloomer, the shy classmate who turns into a star on stage, the dour plodder who becomes a creative genius once their teacher finds and ignites their passion—these kinds of transformations are what teachers live for. (2024, p. 4)

Think of a student in your class who does not yet have a strong math identity. Imagine, for a second, how having a positive math identity could impact this student's future. Challenge yourself to find ways to help this student see themselves as a powerful mathematician by the end of the year. Brainstorm with a colleague some concrete actions you can each take to achieve this goal.

As a group, share out the action steps that were brainstormed. List them on a chart.

Individually, write down three specific actions you will take in the next month to grow your student's mathematical identity.

Let's Try It (10 minutes)

As a teaching community, choose one of the data collection focuses below. Bring this classroom data to next month's meeting to share and discuss.

Take time now to choose a date when you'll gather data from your selected option. Record this task and deadline in your calendar or planner.

OPTION 1: STUDENT INTERVIEWS

Interview at least three students about their mathematical identities. You might use a couple of the following interview questions or create your own:

- What is a math activity you have done at school that you love?

- What are your strengths in math? What are you good at?

- What are you still learning in math?

- Is math important? Why?

- How do you use math in real life?

- I feel happiest in math when _____.

- If I had a wish about math class, it would be _____.

- I love it when my math teacher _____.

- I do my best math work when _____.

OPTION 2: EXIT CARDS

At the end of a math lesson, pose the question "What are you learning about yourself as a mathematician?" Then ask students to write their responses on index cards and hand them to you. Afterward, read and sort the responses in some way that is meaningful to you.

OPTION 3: CLASS DISCUSSION

As an alternative to the writing activity in Option 2, pose the same question and give students a minute of think time. Then invite students to share their thinking in a class discussion. Audio record the discussion with your phone so you can listen later and jot down what students shared.

ADDITIONAL PROFESSIONAL LEARNING EXPERIENCES

This month's additional professional learning experiences relate to the mathematics content strand of measurement. Measurement is a way that people regularly use math in real life, often without consciously thinking about this activity as mathematical. Helping teachers and students to notice ways they apply mathematics in daily routines strengthens their identities as mathematicians.

Math Talks: Measurement Talks

This month's math talk offers you and your colleagues the opportunity to engage in the math habits as you solve a problem involving measurement concepts, think about what it means to measure something, and consider how measurement concepts and skills connect to other areas of the math curriculum.

On a document camera, show the image below. Ask teachers to consider these questions silently and individually for one minute:

- How wide is the sheet of paper?

- How do you know?

Image souce: Istock.com/hocus-focus

online resources ☝ Available as a downloadable resource on the companion website.

Invite team members to share their thinking. Discuss how students might think about and solve this problem. As a group, articulate what it means to measure something.

Discuss the experience:

- What mathematical ideas did you draw on as you thought about this problem?

- What math habits did you engage in?

- At this moment, how would you describe what it means to be a mathematician?

Manipulatives and Models Matter: Create Your Own Manipulative

When we see ourselves as mathematicians, we recognize that mathematics helps us to accomplish specific purposes and makes our lives and the world better. As we engage in mathematics, we often use tools to assist us with mathematical work.

A **mathematical model** becomes a **tool** when a mathematician understands its **mathematical structure** and can therefore use it to solve problems.

Students often have difficulty learning to measure with a ruler because they don't understand a ruler's mathematical structure. One way we can help is to engage students in directly experiencing the structure of this tool by creating their own rulers.

Together with your team, take a few minutes to create your own rulers and then discuss how this experience developed your "mathematical eyesight," your understanding of the mathematical ideas and structure that undergird a ruler as a mathematical tool.

- To prepare: Cut 1×12" strips on a paper cutter out of cardstock, as well as a large quantity of 1" squares in two contrasting colors. You'll also need glue sticks.

- Before creating your own ruler, share your experiences teaching children to use rulers. Discuss misconceptions that commonly surface and why these misconceptions occur.

- Create your own cardstock ruler by gluing alternating colored squares onto a paper strip as shown below. Use your ruler to measure various lengths around the room.

(Continued)

(Continued)

REFLECT

- How would you describe the mathematical structure of a ruler?

- How did the experience of creating a ruler help you to see and understand this structure?

- How does this experience help you to think about what it means to be a mathematician?

- Consider other mathematical tools that play an important role in the team's grade-level curriculum.

  ○ How would you describe the mathematical structure of these tools?

  ○ What are some questions you might ask to determine whether a student understands a tool's mathematical structure?

  ○ What are some ways you can support students in learning about a tool's mathematical structure?

Game Time

For this month, you can use this game-like activity with your students using measurements appropriate to your grade-level curriculum to build students' understanding of measurement and their personal measurement benchmarks. Have students estimate and check lengths, areas, perimeters, volume, angles, time, and so on.

MEASURE AND CHECK

Materials:

Any number of players

Objects from around the room

Rulers (customary or metric)

Pencil

Recording sheet (optional)

Challenge: Be the player whose measurement estimate of an object is closest to the actual measurement.

Game directions:

1. Look around the room. Identify an object that you think is approximately 12" in length.

2. Use your ruler to check. How did you do? Celebrate the colleague whose object was the closest to 12" as the winner!

3. Try again with a different length.

If you want to formalize the game and provide a bit of computation practice, use the recording sheet below and have students play the game with partners. The student with the lowest total score is the winner.

Measure and Check

OBJECT	ESTIMATED MEASUREMENT	ACTUAL MEASUREMENT	DIFFERENCE
Total of all differences for final score:			

Discuss the experience:

• What math habits did you engage in as you played this game?

• How might you adapt this game for use with your students?

• How is your understanding of what it means to be a mathematician growing?

 Available as a downloadable resource on the companion website.

IN YOUR CLASSROOM

Classroom Story

Mr. Sun gathered his kindergarten students on the carpet to debrief from the day's math learning. He had worked with guided math groups on strategies for solving addition problems using ten frames. Other students engaged in learning stations on tasks appropriate to their learning needs. Mr. Sun knew that this 10-minute discussion at the end of math time was an important investment in helping students to solidify their learning, and he was becoming increasingly aware of its value for also building his students' identities as mathematicians.

Mr. Sun opened the discussion with several open-ended questions to help students reflect on their learning.

Mr. Sun: So, what are you learning about addition?

Linza: I like addition. It's fun!

Mr. Sun made a mental note of Linza's enthusiasm, recognizing that emotions play an important role in shaping identities. He knew, however, that Linza's and other students' mathematical identities would grow as they witnessed concrete evidence of their mathematical competence. So, he pressed for deeper thinking.

Mr. Sun: Linza, if you were explaining addition to your little brother, what might you say to him?

Linza: I put some counters on my ten frame, and then I put some more. Then I find out how many I have altogether.

Mr. Sun: Thank you, Linza. Who would like to add on?

Mr. Sun continued to ask questions to help students summarize their learning. He then turned the conversation to their mathematical behaviors.

Mr. Sun: When our principal, Ms. Marquez, comes into our classroom, what does she see and hear that tells her we are mathematicians?

Chris: We talk about our math thinking.

Amelia: We use math tools.

Sandeep: We write number sentences.

Mr. Sun encouraged students to identify their mathematical actions. He ended the class discussion with a final question.

Mr. Sun: What is one way you are proud of yourself as a mathematician today? Tell your turn-and-talk partner.

As Mr. Sun listened into his students' conversations, he realized how much they had grown in just a few months. Mr. Sun took some quick anecdotal notes about students' developing understanding of number and operation concepts and their growing awareness of their mathematical habits.

Anchor Lesson

What Does It Mean to Be a Mathematician?

WHAT?

This lesson will help students to think about what it means to be a mathematician and to see that they are already mathematicians. It will also help them to review the habits of mathematically powerful people that they've learned about in the first half of the school year.

YOU NEED:

Challenging math task

Chart paper

Markers

HOW?

1. Write "I am a mathematician" at the top of an anchor chart. Underneath this statement, write the three habits of mathematically powerful people that students have learned about:

    - "I expect math to make sense."

    - "I love challenging math problems."

    - "I learn from math mistakes."

2. Decide on a mathematics task or problem you will use for this lesson. The problem should challenge students' mathematical thinking and offer opportunities for students to engage in the habits of mathematically powerful people.

    You might use a secret number problem:

    - K–1: I'm thinking of a number between 0 and 20. What might it be?

    - 2–3: I'm thinking of a number. The digit in the hundreds place is larger than the digit in the ones place. The digit in the tens place is larger than the digit in the hundreds place and is odd. What might my number be?

    - 4–5: I'm thinking of a number. If I divide the number by three, there is a remainder of one. If I divide the number by five, there is a remainder of one. What could my number be?

3. Remind students that mathematicians are people who engage in specific habits. Tell students they will work together to solve a math problem. Ask them to notice how they and others use the three math habits listed on the anchor chart.

4. Engage students in the mathematics task or problem you have chosen in partnerships or small groups. Ask students to share how they or others demonstrated the math habits as they thought about and solved the problem. Record their examples on the anchor chart. You may want to have some ideas ready in case students have difficulty coming up with examples.

REFLECT

When we help students notice their use of the math habits and how these habits support their learning, our students feel pride in their growth and are motivated to continue building their math habits.

Use the discussion questions to reflect with students:

- How do the math habits help us learn?

- How can we work to build our math habits?

LET'S DO SOME MATH ACROSS OUR SCHOOL!

We grow our math identities when we notice how we use math in real life. As you and your colleagues did earlier, have your students write true math story problems in which they are the main characters. Their math stories should illuminate the way they use math outside of school.

You might have students trade their stories with classmates to think about and solve. Or you can post students' story problems, inviting class members and others to leave a sticky note telling what they enjoyed or appreciated about the author's mathematical thinking. You might also graph and discuss the different types of mathematical activity shared in the problems. You might do this activity orally in younger grades.

In addition to building your students' math identities and broadening their awareness of how we use math outside of school, this mathematical experience can promote appreciation of students' diverse experiences and strengthen your classroom community.

TEACHING MOVE: USE THE LANGUAGE OF IDENTITY

If we want our students to see themselves as mathematicians, we need to explicitly talk about and model our own mathematical identities and help students to regularly reflect on their own developing mathematical identities. One way we can do this is by incorporating the use of identity language into our teacher talk. We can use phrases such as

- As mathematicians, we _____
- Because we are mathematicians _____
- When we look at this problem as mathematicians _____

REFLECT

- How does our language affect the way students see themselves as mathematicians?

TEACHING MOVE: SCAN FOR NONVERBALS

In their book *Math Recess*, Singh and Brownell (2019) stated, "The shoulders-down posture is an important physical indication that a student is ready—truly ready—to learn mathematics. This is not a metaphor. It's a literal, observable phenomenon in math class. Think about it. A person's emotions are often visible to others, and teachers watch so many of them play out every day. Hunched shoulders indicate frustration, anxiety, even worry, and a relaxed shoulders-down posture—with arms swinging freely—indicates ease of mind and playfulness" (pp. xxvii–xxviii).

As we teach mathematics and observe our student mathematicians at work, we can build the habit of periodically scanning the room for nonverbal indicators of students' emotions. Although we know that productive struggle is important to math learning, students' nonverbal cues can provide us with important data related to their mathematical identities and current capacity to persevere with challenging problems.

REFLECT

- As your students work on challenging math problems, what nonverbal indicators of productive or unproductive struggle are you noticing?

- What are some things you can do when you notice nonverbal indicators of unproductive struggle?

CLASSROOM ROUTINE: GROWING MATH IDENTITIES

At the end of a unit of study, spend a few minutes thinking together with students about how the math they are learning is useful outside of school, right now and in their futures.

You might also post a sentence frame related to math identity once a week and invite students to reflect on their growing identities by completing this frame. Here are some ideas:

(Continued)

(Continued)

- I used to feel _____ about math, but now I feel _____.

- I'm at my best as a mathematician when I _____.

- When I have a challenging math problem to solve, I tell myself _____.

- One way I am growing as a mathematician is _____.

CLASSROOM ROUTINE: EMOTIONS AND MATH

Encourage students to talk about their feelings related to math. Research psychologist Fredrickson (2009) has identified 10 positive emotions that, when experienced regularly, help people to thrive in life: joy, pride, gratitude, amusement, serenity, awe, interest, inspiration, hope, and love.

Create a poster listing these 10 positive emotions. During the close of a math lesson, invite students to share about one positive emotion they experienced during that day's math lesson.

STATION ACTIVITY: THINKING ABOUT MY MATH IDENTITY

We can provide opportunities for students to reflect on their growing math identities during station time. You might use the prompts below or design others aligned with your current math content focus. For instance, if the class is learning about decimal numbers, you might suggest that students write about ways they can use decimals in their future careers.

THINKING ABOUT MY MATH IDENTITY

Choose one of these prompts to write about. Illustrate your writing if you wish. Post your work on a bulletin board for others to read.

- My favorite math memory is _____

- Math is fun when _____

- One of my math superpowers is _____

Read others' work. Give them feedback on a sticky note using this sentence stem:

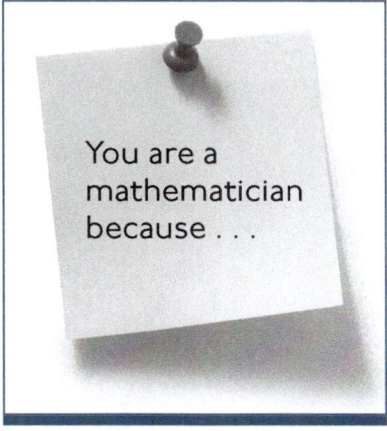

You are a mathematician because . . .

Source: Istock.com/Thammask-Chuenchom

 Available as a downloadable resource on the companion website.

LITERATURE CONNECTIONS

***Count on Me* by Tanco** (2019) (Grades K–2)

The heroine of this story reflects on different ways to see and understand the world and describes how she looks at the world with a mathematical lens.

(Continued)

(Continued)

Activity: What can you see when you look at the world as a mathematician? Take a math walk, looking for examples of mathematics in real life. How many examples can you find?

The World Is Not a Rectangle: A Portrait of Architect Zaha Hadid by Winter (2017) (Grades 3–5)

Iraqi–British architect Zaha Hadid's passion for mathematics led her to design uniquely beautiful buildings and bridges all over the world (see Figure 6.3).

Figure 6.3 • Zaha Hadid's Heydar Aliyev Cultural Center in Baku, Azerbaijan

Image source: Istock.com/benedek

Activity: What is something you're passionate about? How can mathematics help you to explore and use your personal passion?

SPOTLIGHT ON BRAIN SCIENCE

We used to think that only some people were good at math. Students need to know there is scientific proof that all people can understand and learn math. Our brains are wired to make sense of mathematical ideas through multiple modalities and students should tap into all of those modalities in their math thinking and learning (Boaler, 2022).

You can share the following information with students in a mini lesson or a station activity. You might also send this reproducible home along with the Family Newsletter to help families talk about these important ideas.

WE ALL HAVE A MATH BRAIN.

Brain scientists have discovered some important facts about our brains:

- Our brains are constantly changing because we are always learning.

- We can learn math with numbers, and we can also learn math with words, objects, pictures, and from moving and touching things. So when we do math, five different areas of our brain are at work.

- Human beings are born with a math brain. With the right learning experiences, we can understand and do challenging mathematics.

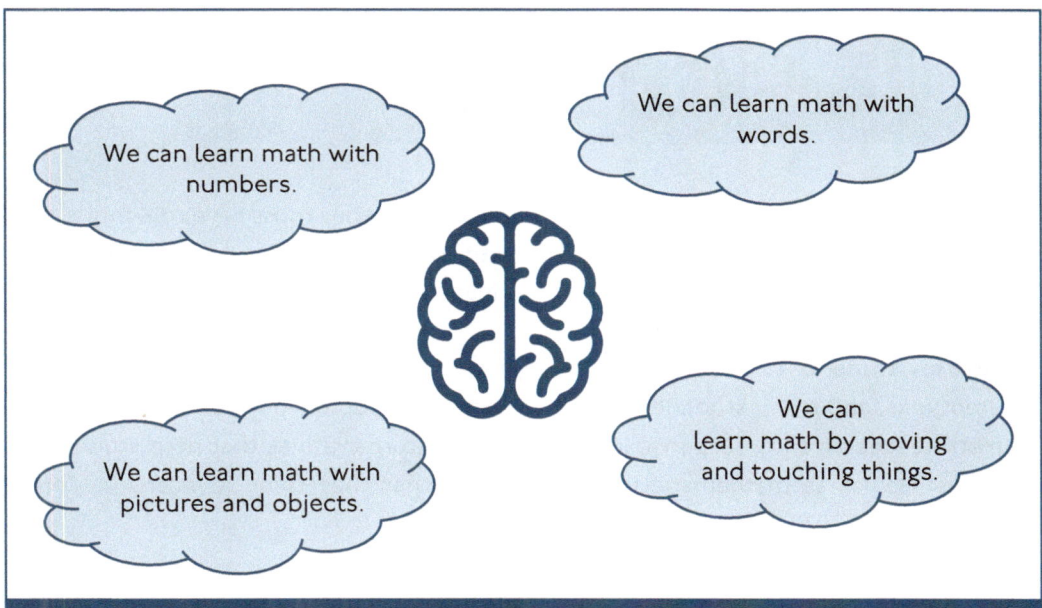

Image source: Istock.com/Pavlo-Stavnichuk

(Continued)

(Continued)

This means that *you* are already a mathematician, and every time you think about or do math, your mathematical abilities become even stronger.

How can knowing these scientific facts help you now and in the future?

Choose one of the following ways to think more about this important information:

1. Write down why these ideas are important to you and how you want to put them to work in your life.

2. Talk to a friend about these ideas. Decide together how you will put them to work.

3. Share these ideas with someone at home. Tell them why these ideas are important to you and how you plan to use them.

 Available as a downloadable resource on the companion website.

SPOTLIGHT ON EQUITY: MATH ROLE MODELS

If we want students to see themselves as mathematicians, they must have role models who look like them and that they can identify with. Kane, author of the Edutopia blog post "How to Help Students See Themselves as Mathematicians," stated:

> I want my students to believe they can be mathematicians. But math is plagued by stereotypes, and many students think math isn't for people who look like them. As a math teacher, it's my responsibility to offer counternarratives that help students see themselves as mathematicians and expand what they think it means to "do mathematics." (2020; https://edut.to/4alc37H)

TRY IT

What are some ways we can give our students access to strong mathematician role models who share their identity markers (race, gender, culture, language, religion, etc.)?

- Are there colleagues in your school with identity markers different from yours who would be willing to share their math stories with your class?

- Can you arrange for former students with positive math identities who are now in middle and high school to visit your class?

- What biographies of diverse mathematicians are available in the library? Check out the list of such biographies in the online resources.

MATHEMATICAL ME: STUDENT JOURNAL AND PORTFOLIO

STUDENT JOURNALS:

Have students respond to the following questions:

- What does it mean to be a mathematician?
- How do I know I am a mathematician?

Older students can write their responses in their Mathematical Me journals. Younger students can draw a picture showing their response, or you might gather responses during a class discussion or quick one-on-one interviews.

Review students' responses to monitor students' mathematical identities and agency. You might summarize this data by tallying or graphing the different ideas students offer.

Be sure to share and discuss this data with your class. Ask, "What does this data tell us?" and "What are some things we can do to continue growing as mathematicians?"

STUDENT PORTFOLIOS:\

Have students choose a piece of work that shows their math thinking. On a sticky note, have them write the date and complete this sentence frame:

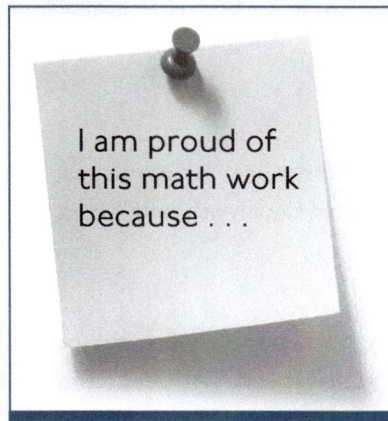

I am proud of this math work because . . .

Source: Istock.com/Thammask-Chuenchom

REVISITING THE ATTITUDE SURVEY

Midway through the year is a great time for students to revisit the Attitude Survey that was taken in August. Administer the survey and guide students in reflecting on and comparing their results from the beginning of the year until now. On a sticky note, have them complete the following sentence frame:

I have grown since the beginning of the year by _____

Be sure to save both surveys and the sticky note in students' portfolios for the end of the year.

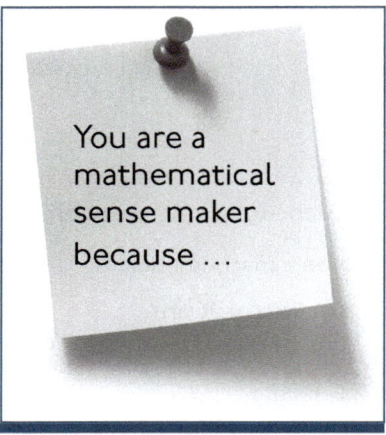

You are a mathematical sense maker because …

Source: Istock.com/Thammask-Chuenchom

FAMILY NEWSLETTER: WE SEE OURSELVES AS MATHEMATICIANS!

WHAT WE'RE LEARNING

This month we're learning about our math identities, our internal beliefs about our ability to learn math and about how math helps us.

WHY IT'S IMPORTANT

We used to think that only some people were good at math. Fortunately, this belief has been disproven. We have solid research evidence that all human beings are capable of learning high-level mathematics. Research also shows that when children believe they are capable of understanding math and that math is useful and interesting, they are more successful and better equipped for future academic and career opportunities.

HOW YOU CAN HELP

As you engage in daily routines, talk with your child about the math you are using:

- This recipe calls for three-fourths cup of flour. Three-fourths cup is less than one cup. I could measure this amount with three one-fourth cups or with one-half cup and one-fourth cup.

- Our car holds 15 gallons of gas. Today gas costs $___ a gallon. Let's estimate how much it will cost to fill up our gas tank.

- How much time does it take us to get ready in the morning, and what time do we need to wake up?

Share ways that you and others use mathematics at work. Help your child understand that everyone is a mathematician and that math helps us in all aspects of our lives.

Listen for and discuss examples of math in the news. Help your child grow the habit of actively looking and listening for all the ways math benefits us.

Many adults have had negative past experiences with math. Be careful not to pass negative feelings about math onto your child. When you talk about math, focus on the ways that math helps you and is important.

 Available as a downloadable resource on the companion website.

CHECKING IN ON OUR LEARNING: WE SEE OURSELVES AS MATHEMATICIANS!

- What does it mean to be a mathematician?

- How do I know I am a mathematician?

This month's learning focus, mathematical identity, is integral to effective mathematics teaching and learning. Our math identities and our students' math identities are braided into our experiences as math learners. In a powerful math community, teachers and leaders hold themselves account-able for building all students' math identities alongside their mathematical understandings and skills.

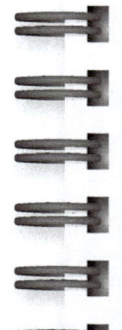

MATHEMATICAL ME: EDUCATOR JOURNAL

- How has this month's learning focus supported your students' mathematical growth?

- How has this month's learning focus supported your growth as a math teacher?

- How has this month's learning focus supported your school's growth as a powerful math community?

LOOKING AHEAD

In January, we will explore how talking about math can strengthen our mathematical identities and our math learning.

Questions for Reflection

- As an adult math user and learner, when do you find yourself talking about mathematics with others both in school and outside of school?

- How do these mathematical conversations support you and others?

CHAPTER SEVEN

January
We Talk About Math!

ESSENTIAL QUESTIONS

- Why is it important to talk about my math thinking?

- How can I get better at talking about my math thinking?

THIS MONTH'S FOCUS

In a powerful math community, students and educators talk about their mathematical thinking to support their own and others' reasoning and sense-making. They know that mathematical discourse involves both speaking and listening, as well as nonlinguistic forms of communication, and that this exchange of mathematical ideas helps the community co-construct mathematical knowledge and build mathematical proficiencies.

Educators believe that students can and need to learn how to communicate about and with mathematics. They hold themselves responsible for developing students' understanding and use of math vocabulary and their math communication skills.

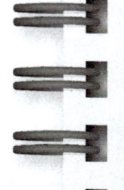

MATHEMATICAL ME: EDUCATOR JOURNAL

Take five minutes to respond to the essential questions. You will have a chance to reflect on these same questions and to have students respond to these questions at the end of the month.

ON YOUR OWN

LET'S DO SOME MATH! MATH RIDDLES

Image source: lstock.com/stick-figures-com

Solve one of these riddles. (When you're done, turn to the end of this chapter for answers.)

1. I'm thinking of four even numbers.

    Each number has two digits.

    Each number is double the previous number.

The product of the digits in each number is 0.

What are my four numbers?

2. I have eight coins in my pocket.

   I have three different types of coins.

   Three fourths of the coins are silver in color.

   The total value of all the coins is an odd number.

   I have less than $1.00.

   What coins do I have?

3. My students wanted to guess my age, so I gave them these math clues:

   My age is divisible by two and seven.

   The sum of the digits in my age is less than 10.

   Write a final clue for this riddle so there is only one possible solution.

Now, write a math riddle of your own related to the math content you are currently teaching. Save this riddle to share with colleagues. Consider how you engaged in the habits of mathematically powerful people as you solved a riddle and wrote your own riddle.

 Available as a downloadable resource on the companion website.

Why This Focus

Sue's Math Story:

My husband Gerry loves to tell corny jokes that make people laugh or even just groan and roll their eyes. And because he supports my passions, Gerry is always on the lookout for math jokes to share with me.

When I check my email in the morning, a couple of math jokes are often waiting for me.

What tool should every student have?

Multi-pliers.

What did the calculator say to the mathematician?

You can count on me.

Who is the king of all the math tools?

The ruler.

Over dinner, Gerry will frequently try out new math jokes he's found.

> Where do mathematicians eat their dinner?
>
> *At the multiplication table.*

> Why did the two fours skip dinner?
> *They already eight.*

Gerry is convinced that telling jokes and making people laugh makes the world a better place. He encourages me to share his math jokes with the teachers and students I work with.

> Why did the obtuse triangle lose every argument?
>
> *He was never right.*

> Why shouldn't you argue with a circle?
>
> *It is pointless.*

> Which angle is the most adorable?
>
> *An acute angle*

> Here's my all-time favorite:
>
> Why did the math teacher only solve subtraction problems?
>
> *She wanted to make a difference.*

These math jokes were inspired by *Super Silly Math Jokes for Kids* by Mashup Math (2023). Gerry believes that amusement should be a reliable part of daily life and that jokes, especially corny jokes, help others to discover a smile. Riddles and jokes make us think. They cause us to smile. They offer an element of surprise that engages and tickles us. Riddles and jokes involve the playful use of language, and language can bring people together.

This chapter is about the use of language within a powerful mathematical community, how language is used to share ideas and the types of mathematical ideas that are valued and therefore shared. It is also about who does the sharing in a math community, whose voices participate in the community's exchange of ideas. It is about **mathematical discourse** or "the purposeful exchange of [mathematical] ideas through verbal, written, and visual communication" (Huinker & Bill, 2017, p. 178).

A math community is powerful when all community members play an active role in its mathematical discourse. The exchange of mathematical ideas positions students as math thinkers and doers, as individuals who are powerful because of the mathematical understandings they possess and contribute to the community's shared knowledge.

This type of math community is different from the classrooms most of us experienced as students in school where the teacher's voice was predominant and student talk amounted to answering questions with a single correct answer. Making room for student thinking and student voices in the math classroom requires that we, as teachers, regularly quiet our own voices. Education thought leader Kohn (2013) stated, "Terrific teachers have teeth marks on their tongues."

Because this type of teaching differs from how we were taught, we must believe in its importance and invest time learning how to make it happen in our classrooms. Reinhart described how he learned, over time, to value and build a shared discourse of ideas in his math classroom:

> Eventually, I concluded that if my students were to ever really learn mathematics, they would have to do the explaining, and I, the listening. My definition of a good teacher has since changed from "one who explains things so well that students understand" to "one who gets students to explain things so well that they can be understood." (2000, p. 478)

When our students talk about their mathematical thinking, when our classroom communities share and co-construct mathematical knowledge, we all learn and are enriched by the ongoing exchange of mathematical ideas.

Teachers as "Voice Coaches"

Explaining and discussing their math thinking is more difficult for some students than others. Our students come to us with a beautiful array of learning strengths and differences. It is no secret that students with language processing differences, students learning math in a non-native language, students with cultural norms that discourage speaking up, and students who are by nature introverted must all work harder at talking about their math thinking. We need to meet our learners where they are while expecting and supporting growth (Crespo et al., 2018). We need to help each of our students find their "math voice."

Communication supports learning and academic success and is, therefore, important for all students. Communicating about their math thinking strengthens students' mathematical understanding and helps learning move into long-term memory (Chapin et al., 2022). As students practice communicating about their learning, they acquire a habit that will support future learning. The ability to share their mathematical thinking also positions students as valued members of a learning community.

> **A math community is powerful when all community members play an active role in its mathematical discourse.**

Teachers should expect to scaffold students' use of language just as we scaffold the development of other skills. According to Lambert (2024), "We should not remove or simplify language but instead provide access to it" (p. 135). This critical scaffolding can take many forms, including:

- Use of manipulatives, visuals, gestures, sentence frames, and word banks to support communication

- Opportunities to practice using language, including partner work and rehearsals

- Use of native language for multilingual learners

- One-to-one coaching

At the same time, we also want to look beyond language-based communication to the broader concept of classroom discourse, all of the ways that mathematical thinking can be shared. The concept of discourse overlaps with the topic of mathematical representation, which is considered in the next chapter. We want all students to be able to share their math thinking in multiple ways, including nonlinguistically, and to value multimodal communication.

Mathematical Discourse

Mathematical discourse promotes student learning. Chapin et al. (2022) identified five reasons that student talk is critical to math learning:

- Talk can reveal understanding and misunderstanding

- Talk supports robust learning by boosting memory

- Talk supports deeper reasoning

- Talk supports language development

- Talk supports development of social skills (p. xv)

Questions for Reflection:

- Think of examples of how you've seen student talk support each of these aspects of learning?

- What structures do you use to support student talk for these purposes?

Productive math discourse doesn't happen automatically. When students are not yet skilled in talking about their math thinking, they have difficulty engaging in mathematical work with others and the class has difficulty functioning as a learning community. Max Ray, author of *Powerful Problem Solving*, stated:

Not feeling comfortable putting mathematical ideas into words or writing is frightening to students and makes it hard for them to want to collaborate, ask questions, share ideas, or articulate things about math besides the steps they already know how to do. (2013, p. 9)

Project Challenge

From 1998 through 2002, Chapin et al. (2022) engaged in a research study called Project Challenge designed to find ways of increasing the number of underserved students qualifying for gifted and talented programs in urban school districts.

Researchers helped teachers learn strategies for developing students' abilities to talk about their mathematical thinking. Within a year, Chapin et al. noticed that "students' reasoning had become more complex, more sophisticated, and more recognizably mathematical. Students were better able to give clear explanations for the problem solutions, their use of language became more precise, and their communication skills improved noticeably" (2022, p. 321). This growth was also evident in students' mathematical confidence and achievement test scores. In short, teachers' intentional development of students' communication skills had a strong impact on their learning, mathematical identities, and agency.

The talk facilitation moves in Table 7.1 are based on the techniques used in the Project Challenge research and the work of math education leaders Huinker and Bill (2017) and Ball (TeachingWorks, 2023). When teachers across a school become skilled in the strategic use of these talk facilitation moves, students' communication skills and mathematics learning can improve dramatically. In fact, when Holly and I work with schools and districts to improve math instruction, one of our first recommendations is to support teachers in building their questioning skills and toolkits of talk facilitation moves.

> When teachers across a school become skilled in the strategic use of these talk facilitation moves, students' communication skills and mathematics learning can improve dramatically.

Talk Facilitation Moves for Teachers

The skills of productive math talk, like all skills, can be learned. Whole-class discussions, small-group collaboration, and partner work are all settings where students can practice and grow their abilities to talk about their math thinking. Within each of these classroom settings, teachers can help students build proficiency in talking about their math thinking and listening to others talk about their math thinking through the use of talk facilitation moves. Specifically, teachers need command of four types of talk facilitation moves to use flexibly in conversational settings with students:

- Eliciting moves
- Clarifying moves
- Probing moves
- Orienting moves

Table 7.1 offers examples of how a teacher might use these four moves in class discussion about the age riddle.

My students wanted to guess my age so I gave them the following math clues:

My age is divisible by two and seven.

The sum of the digits in my age is less than 10.

Write a final clue for this riddle so there is only one possible solution.

Table 7.1 • Four Types of Talk Facilitation Moves

MOVE	PURPOSE	EXAMPLES OF TEACHER MOVES
Eliciting	Move that supports students in initially articulating their thinking	Based on these clues, what are some things you know about my age?
Clarifying	Follow-up move to support students in clarifying or refining their thinking	You said that 14 is divisible by two and seven. What does that mean? You said that I couldn't be 56 years old. Why do you think that?
Probing	Follow-up move to support students in thinking more deeply about an idea	So, you've identified some ages that would be possible based on these clues. What else might we need to consider in solving this riddle? How could we create a final clue based on that idea?
Orienting	Move that supports students in thinking about and responding to another's idea	What do you think of _____'s idea? Does it work? How do you know? What are some other possibilities for a final clue?

What types of talk facilitation moves do you currently use to support student thinking and communication? Are there moves that are your "go-to" moves? Are there moves that you use infrequently?

Audio record yourself facilitating a five-minute classroom discussion using an app on your phone. As you listen to the recording, jot down some notes about the types of talk facilitation moves you used to support student thinking and communication. Save these notes to share with colleagues in the Together With Your Teaching Community learning activities.

Tapping Into Your Experience

Chapin et al. (2022) identified two essential norms for a classroom community with productive math discourse:

- **Respectful Discourse**: Talk is respectful when each person's ideas are taken seriously; no one is ridiculed or insulted, and no one is ignored or browbeaten.

- **Equitable Participation**: Participation is equitable when each person has a fair chance to ask questions, make statements, and express his or her ideas. Academically productive talk is not just for the most vocal or the most talented students. (p. 69)

Think of a classroom you were part of that was defined by equitable participation and respectful discourse. Now, think about a different classroom in which these two norms were not in place.

- How did the presence or absence of these norms affect your learning?

- How did they affect the ways these groups functioned?

- How did they contribute to group members' feelings of belonging?

- How were these norms built and reinforced by group leaders and members?

On a scale of one to five, how would you rate the current presence of these norms in your math classroom?

Respectful Discourse

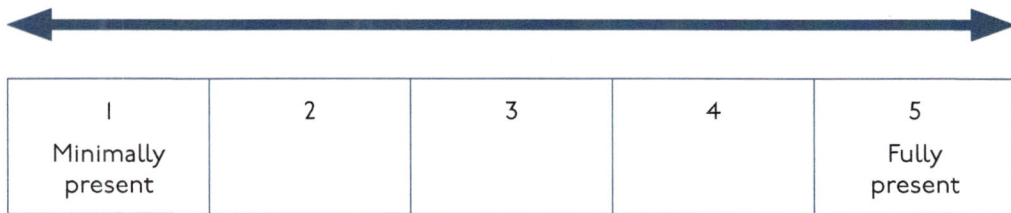

1	2	3	4	5
Minimally present				Fully present

Equitable Participation

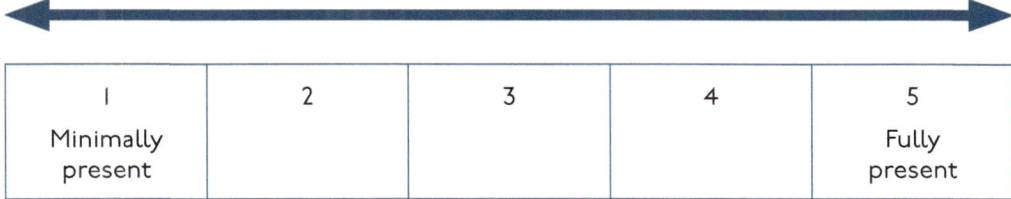

1	2	3	4	5
Minimally present				Fully present

What are some things you can do to strengthen these norms in your classroom?

How are respectful discourse and equitable participation enacted across your school? How might you and your colleagues work together to grow these norms as a feature of your school culture?

Productive math talk does not happen by accident. We help students develop proficiency in talking about mathematical ideas through our strategic use of talk facilitation moves. We also support this important aspect of student learning as we structure and plan opportunities for students to practice the skills of discourse in various classroom contexts.

Here are some ideas for instructional design in various classroom contexts to develop students' discourse skills.

Whole-Class Math Talk

Class discussions of complex mathematics problems and ideas are a keystone learning experience in a powerful math community. These discussions involve much more than the traditional routine of students providing short answers to questions asked by the teacher. Instead, students are expected to explain and justify their mathematical thinking and respond to others' reasoning (Chapin et al., 2022; National Council of Teachers of Mathematics [NCTM], 2020). Huinker and Bill described this whole-class discourse:

> The essence of whole-class discussions is making student thinking visible and public for discussion and examination. . . . Engaging students in class discussions that analyze and compare students' thinking and reasoning across a variety of solution paths validates the contributions of each learner and supports the classroom as a mathematical learning community. (2017, p. 209)

Teachers need to prepare for class discussions as a consistent part of their instructional planning routines. Kobett and Karp (2020) suggested that teachers "plan at least three strategic moments in every lesson for students to communicate their ideas with each other" (p. 50). As teachers plan lessons, they should consistently craft open-ended questions designed to focus students' attention on the big mathematical ideas of the lesson. Teachers can pose these questions as students work on the problem and during the class discussion at the end of the lesson.

Small-Group Math Talk

Collaborative learning work with thinking-rich mathematics problems in partnerships and small groups offers another important opportunity for students to practice the skills of mathematical discourse. According to Hattie et al.:

> Students learn a lot more language when they are required to produce language. Mathematics is a language, foreign to some and familiar to others. One of the best ways to apprentice students into the language of mathematics, which then facilitates their mathematical thinking and reasoning, is to have them collaborate with their peers in solving complex, rich tasks. (2017, p. 153)

Liljedahl (2021) recommended regularly changing partnerships and collaborative groups and forming new groupings using randomized methods that allow students to see the randomization taking place. Liljedahl's research demonstrated the following benefits to the use of frequently changed, visibly randomized groups:

- Willingness to Collaborate

- Elimination of Social Barriers

- Increased Knowledge Mobility

- Increased Enthusiasm for Mathematics Learning

- Reduced Social Stress (2021, pp. 46–49)

To randomize groups easily, you might have each student select a playing card as they enter the classroom, creating groups of students with the same type of card. A variety of technology-based randomizers are also available. Liljedahl stated that the optimal group size for students in K–2 is two students and that groups of three are best for students in third grade and above.

Math Class Can Make the World a Better Place

In his book *Dear Citizen Math*, Karim Ani asserted that

> Public education exists for two purposes: to help students develop the knowledge and skills they'll need to succeed in their individual lives, and to create opportunities for children to learn together in order to strengthen our social fabric and ensure a healthy democracy. As math educators, we have an important role to play in each of these. At this fragile moment in our national history, though—a moment in which our discourse is imperiled by irrationality, incivility, and a lack of critical thought—it is our social contribution that's more urgent than ever. (2021, pp. 106–107)

Ani (2021) maintained that as we teach math, we also possess the precious opportunity to gift our students with discourse skills that are sorely needed in society. As students learn to discuss mathematical ideas, they develop social and emotional intelligence. They are practicing respect, patience, and kindness. Students are developing the habit of listening to understand another person's point of view and the ability to be clear and precise in their explanations (Chapin et al., 2022). When students learn to talk respectfully about complex ideas, they are better equipped for life and the world benefits from these proficiencies and mindsets.

Math class, when done right, can create a more respectful and equitable society. Math class, when done right, can change the world.

TOGETHER WITH YOUR TEACHING COMMUNITY

Since We Met Last (10 minutes)

In December, you and your colleagues collected classroom data about students' mathematical identities.

Use the following protocol to share and reflect on this data with your math teaching community:

- Share your data with a colleague. (5 minutes)

- As a group, discuss what you learned from the data and how these insights can help you to support your students' growth as mathematicians. (5 minutes)

Let's Do Some Math Together! (10 minutes)

Talk with your team about your experience solving a riddle.

Discuss:

- How did it feel to engage with and solve the riddle?

- What math habits did you use?

Share the riddles that you created with each other.

Discuss:

- How would students benefit from engaging with and solving these riddles?

- How might you use these riddles in the classroom?

Building Our Expertise (30 minutes)

Effective math communication requires the use of precise mathematical vocabulary. Frequently individual teachers are left on their own to decide which vocabulary to introduce to students at each grade level. The result is that, across classrooms and grade levels, teachers unknowingly expose students to different names for the same concept. Students in one class are taught to call ◊ a "diamond." Next year, their new teacher calls this same shape a "rhombus." Or students learn to read 18 – 9 as "eighteen take away nine" rather than the more precise "eighteen minus nine." As they get older, these students are puzzled when they hear that "take away" is only one kind of subtraction. These inconsistent uses of math vocabulary create confusion about the concepts the words represent and erode students' confidence in their ability to communicate mathematically. According to Karp et al.:

> When the language and symbols students use are constantly shifting with each teacher or grade level, communication about one's thinking becomes scattered and muddy rather than cogent, cohesive, and connected. Teaching precise mathematical terminology and using those terms consistently shape students' ability to express their mathematical ideas. (2021, p. 19)

We all tend to default to using math vocabulary we learned as students. Unfortunately, some of these math terms have "expired." They were in common use in the past, but math educators have since recognized that these words do not accurately describe mathematical ideas and therefore do not support students' long-term mathematical growth (Karp et al., 2021).

(Continued)

(Continued)

Here are some examples of math words that have expired and their replacements:

WORD USED PREVIOUSLY	RECOMMENDED REPLACEMENT WORD	EXPLANATION
borrow (in subtraction)	regroup	When we subtract 35 – 7, we don't "borrow" a ten with the intention of giving it back. Instead, we "regroup" one ten as ten ones.
reduce (a fraction)	simplify or write in lowest terms	When we rewrite $\frac{6}{8}$ as $\frac{3}{4}$, we are not "reducing" $\frac{6}{8}$ or making it smaller. Rather, we are simplifying $\frac{6}{8}$ or rewriting it in lowest terms.

Karp et al. (2021) recommended that teachers engage in a collaborative vocabulary alignment process within grade-level teams and across the whole school involving the following three steps:

- Examine the vocabulary currently in use within individual classrooms and at different grade levels.

- Decide together on the vocabulary students will be exposed to at each grade level. These vocabulary lists should reflect grade-level curriculum standards and include only words that accurately and precisely describe mathematical ideas.

- Commit to using their words in instruction and supporting students in using these words in the mathematical explanations.

Begin this process by looking at an upcoming unit of study with your team.

- List the vocabulary that students need to understand and use to communicate effectively about the mathematics they will explore in this unit.

- Check to make sure the words you identify align with your curriculum standards and that they accurately describe mathematical relationships.

- Discuss how you will introduce and develop this vocabulary. Plan anchor charts and sentence frames. Consider how you will monitor growth in student use of the targeted vocabulary. Table 7.2 shows an easy means of gathering data related to students' vocabulary use.

Table 7.2 • Tally Chart for Student Vocabulary Use

STUDENT	ADDEND	PLUS	EQUALS	SUM
Idunn				
Taniya				
Oriane				
Jochen				

- Schedule times to repeat this process for other units of study and to share vocabulary lists across grade levels to ensure consistency in students' vocabulary development across the school.

Let's Try It (10 minutes)

Together with your teaching community, collect some data about your current use of the talk facilitation moves and then work together as a team to build this important teacher skill set across classrooms in your team or school.

1. Share what you learned from listening to the audio recording of your facilitation of a class discussion from On Your Own (see p. 208). List examples of the four types of talk facilitation moves that you and our colleagues used on chart paper or a Google document. Identify moves that you use frequently and moves that are used less frequently.

2. Individually or as a team, decide on a goal to build your proficiency with the talk facilitation moves. You might, for instance, work toward asking more probing questions to deepen students' thinking and mathematical understanding. Or you could focus on using more orienting moves to help students learn how to listen thoughtfully to their peers' ideas and then use these ideas to co-construct mathematical understanding.

3. Once you decide on your learning goal, jot down several questions or language frames from the examples below on an index card or sticky note to keep with you during instruction. Try out these moves as you facilitate math talk and notice how they impact student thinking and communication.

(Continued)

(Continued)

4. If possible, arrange with a colleague to observe class discussions in each other's classrooms to gather data on your use of these moves and their impact. If this is not possible, audio record yourself again and reflect on your learning.

MOVE	PURPOSE	QUESTIONS AND SENTENCE FRAMES
Eliciting	Move that supports students in initially articulating their thinking	• Tell us about your thinking. • How did you think about this problem? • Please share your strategy with the class.
Clarifying	Follow-up move to support students in clarifying or refining their thinking	• So you're saying _____ • Say more about that idea. • I'm not sure I understand. Can you explain it to me in a different way? • Turn and talk with a partner about this idea. What did you and your partner talk about?
Probing	Follow-up move to support students in thinking more deeply about an idea	• How do you know that? • How did you get that answer? • Why does that strategy work? • Will it always work? • Can you prove that to us? • How might you justify that idea?
Orienting	Move that supports students in thinking about and responding to another's idea	• Who can restate that idea in your own words? • What does _____ mean when they say _____? • Who can add on to that explanation? • Do you agree or disagree with _____? Why? • Let's look at these two strategies. How are they alike? How are they different? • Turn and talk with a partner about _____. Who can tell the class what your partner said?

ADDITIONAL PROFESSIONAL LEARNING EXPERIENCES

This month's additional professional learning experiences use the manipulative model of color tiles. Color tiles are one-inch square tiles in four colors: red, blue, green, and yellow. They can be used to represent a wide variety of math concepts, including number, operations, patterns, even and odd numbers, arrays, area and perimeter, and fractional numbers. They are therefore useful across the elementary grades.

Math Talks: What's in the Bags?

This month's math talk is a fun, logical reasoning problem that demonstrates the benefits of talking about our mathematical thinking.

Before the session, gather three identical paper lunch bags. Place two red color tiles in one bag. With a marker, write a large Y (for yellow) on the front of the bag. Place two yellow tiles in the second bag and write a large R + Y (for red and yellow) on the front of the bag. Place a red and a yellow tile in the final bag and label it R (for red).

- Tell group members that you have a mystery that they can solve by talking and thinking together.

- Show the three closed bags and explain that one bag contains only red tiles, one bag contains only yellow tiles, and that one bag contains both red and yellow tiles. State that each of the bags has been mislabeled.

- Explain that group members need to figure out what color tiles are in each bag without opening the bags and looking inside. State that they can only reach into one bag and pull one tile out of that bag. Challenge group members to figure out which bag they should pull a tile from.

Discuss the experience:

- What math habits did you engage in as you solved this mystery?

- What talk facilitation moves did the facilitator use to support thinking and problem-solving?

Inspired by the Three Sack Problem in *About Teaching Mathematics* (Burns, 2007, p. 136)

Manipulatives and Models Matter: Color Tile Riddles

Color tile riddles offer a great opportunity for students to practice talking about ideas in a problem-solving setting and listening to other's ideas. Try this with your math teaching community.

Together with a partner, solve this riddle. Discuss each of the clues and represent your thinking with color tiles. As you work together, notice how talk supports your understanding of the clues and your thinking about the solution to the riddle.

Clues:

- I have 12 color tiles in the bag.

- There are four colors of tiles.

- There are an even number of red tiles.

- There are twice as many blue tiles as green tiles.

- One twelfth of the tiles are yellow.

Write a color tile riddle of your own that could be solved by your students. Decide on the number and colors of tiles you will use. Put them in a lunch bag. Then write three to five clues that allow your students to figure out what color tiles you have in your bag. You can incorporate math vocabulary into your clues to reinforce concepts that are a part of your grade-level curriculum.

After students have solved several riddles together as a class, you can teach students to write their own color tile riddles. Writing and solving color tile riddles works well as a station activity.

REFLECT

- How does the color tile riddles experience allow students to practice their math talk?

Inspired by Riddles with Tiles (Burns, 2022, p. 276)

Game Time

This month's game links the Math Talk and Manipulatives and Models Matter sections by using color tiles to explore addition and multiplication concepts and build fact fluency.

FILL THE GRID

Materials:

2 players

Two dice

Color tiles

Game board for each player — 7×10 grid of one-inch squares

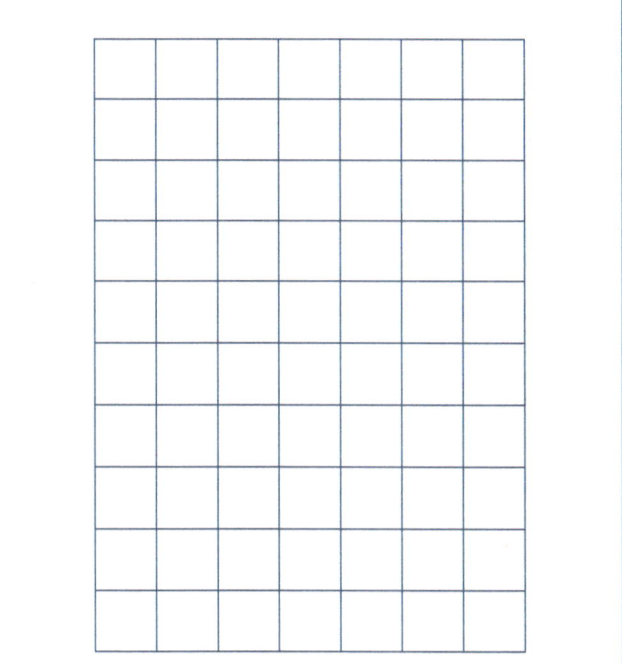

Challenge: Be the first player to completely fill in your grid.

(Continued)

(Continued)

Game directions:

FILL THE GRID (ADDITION)

1. On your turn, roll the two dice. Add the two numbers that you roll and write an addition equation showing the two addends and the sum.

2. Take the number of color tiles equal to your sum and place them in a horizontal or vertical line on your game board. If your sum does not fit in a horizontal or vertical line on your game board, you may use the number of color tiles equal to one of your addends.

3. Play until you or your partner have completely filled the game board.

Example of a Player's Game Board

Turn one: 3 + 4 = 7

Turn two: 5 + 3 = 8

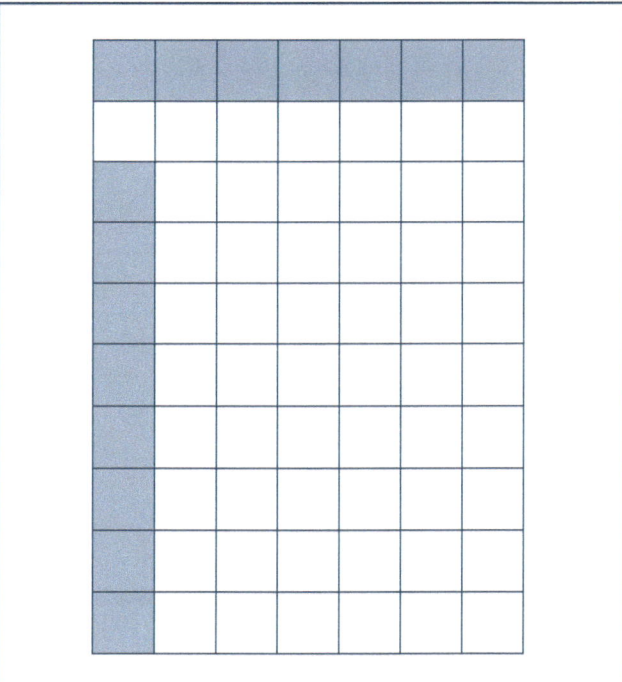

FILL THE GRID (MULTIPLICATION)

1. On your turn, roll the two dice. Multiply the numbers that you roll and write a multiplication equation showing the two factors and the product.

2. Place the number of color tiles equal to your product in an array on your game board. The first factor in your equation is the number of rows in your array, and the second factor is the number of columns.

3. If your array doesn't fit on your game board, you may break it into equal rows or columns and place these rows or columns on your game board. If these rows or columns don't fit on your array, you lose your turn.

4. Play until you or your partner have completely filled the game board.

Example of a Player's Game Board

Turn one: $3 \times 2 = 6$

Turn two: $5 \times 4 = 20$

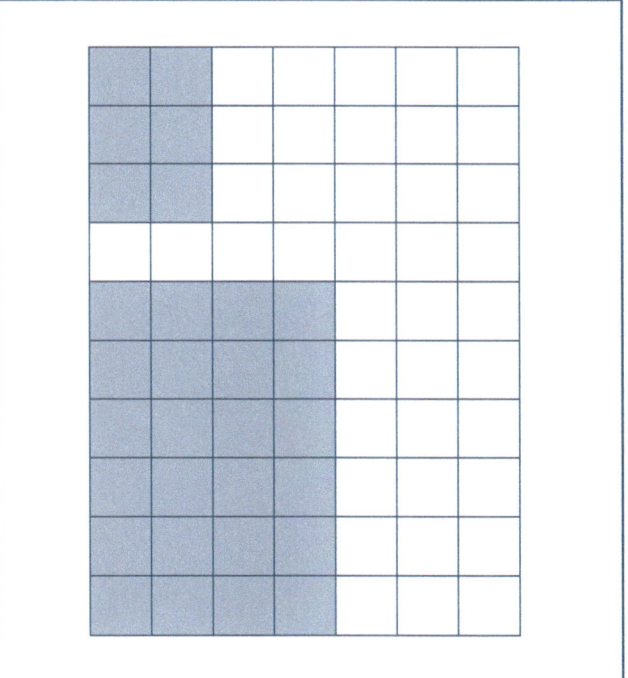

(Continued)

(Continued)

Discuss the experience:

- What math habits did you use when playing the game?

- How might you incorporate math talk into gameplay?

- What mathematical vocabulary would you expect students to use?

 Available as a downloadable resource on the companion website.

IN YOUR CLASSROOM

First, imagine yourself asking questions like the following during a class discussion, as students work in collaborative groups, or as you confer with an individual student. What kinds of thinking do these questions spark for students? When questions like these are used regularly in the math classroom, what habits of mathematical thinking are cultivated?

- Does this idea make sense? Why or why not?

- What's another way we could think about this?

- How is this strategy similar to and different from _____'s strategy?

- Do you agree or disagree, and why?

- How might you prove your idea to someone else?

- What might happen if_____?

Now, visualize your students making statements like these in a class discussion, as they talk to a peer about their mathematical work, possibly even as they think out loud to themselves. How do these statements mirror the teacher questions? What do statements like these reveal about students' math habits, their readiness to learn, their mathematical identities and agency?

- This idea makes sense to me because _____.

- Another way to think about this is _____.

- This strategy is similar to _____.

- I disagree with this idea because _____.

- I'm not sure I understand. Can you explain why _____?

- What might happen if _____?

We can help our students grow as mathematicians by modeling the habits of mathematical thinking in our questions and by prompting them to use specific verbal moves as they talk about their math thinking: explaining, justifying, connecting, critiquing, conjecturing, and questioning. When students learn what these moves sound like and use them when talking with others and to themselves, reasoning and sense-making, as well as classroom discourse, are strengthened, supporting all students' math learning.

Anchor Lesson

Day 1: What Are Some Ways We Can Talk About Our Math Thinking?

WHAT?

This lesson will build student understanding of how talk supports math learning. It will empower students with specific vocabulary and verbal skills they can use to think about their own math talk habits and the class's patterns of discourse.

YOU NEED:

Problems for number talks

Chart paper

Markers

HOW?

1. Tell students that when we talk about our math thinking and listen to others share their math thinking, it helps us to understand math ideas better. Talking about our math thinking also helps us to remember ideas better.

2. Video record your students engaged in a short classroom discussion of a math problem. A number talk is ideal for this purpose because of its brevity (see explanation of number talks in Chapter 5—November).

3. Tell students they will watch the video and, as they do, they should notice how talk helps students to learn. Show the video and debrief by asking students to share things they noticed.

4. Display the anchor chart with the talking-to-learn moves and definitions.

    *   Explaining and justifying—sharing an idea and giving a reason or providing evidence

    *   Connecting—adding onto an idea or making a mathematical connection

    *   Conjecturing—suggesting a mathematical idea or strategy to try

    *   Critiquing—agreeing or disagreeing respectfully with an idea

    *   Questioning—asking a question of another student, the teacher, or the class

Watch the video a second time. This time, pause the video to point out examples of students using the talking-to-learn moves. If there are no examples of some of the moves in the video, model these moves yourself to illustrate what they sound like. Ask students how the talking-to-learn moves support learning.

Note: With younger students, you might focus on just one talking-to-learn move at a time and/or simplify the name of the move (e.g., agreeing or disagreeing instead of critiquing).

REFLECT

When we give students names for specific communication skills they can strengthen through practice, we empower them to notice their use of these skills and monitor their growth.

Use the following discussion questions to reflect with students:

*   Which of the talking-to-learn moves are easy for you to use?

*   What is one talking-to-learn move you might want to strengthen? Why?

Day 2: Using the Talking-to-Learn Moves

WHAT?

This lesson allows students to practice using the talking-to-learn moves to build their confidence in talking about their math thinking and their appreciation for the importance of mathematical discourse.

YOU NEED:

Problems for number talks

Chart paper

Markers

HOW?

1. Prepare anchor charts with language stems for each talking-to-learn move. You may use these language stems or create other stems appropriate for your class. It's fine to start with a single stem for each of the talking-to-learn moves to scaffold student understanding of these moves.

Explaining and justifying

- I think _____ because _____.
- This makes sense to me because _____.
- This seems reasonable to me because _____.
- I chose this method because _____.

Connecting

- This idea connects with _____ because _____.
- This is similar to/different from _____ because _____ .
- Another way to think about this is _____.
- Another strategy is _____.

Conjecturing

- I think we should try _____ because _____.
- I think the pattern is _____.
- I'm noticing _____ and wondering _____.

Critiquing

- I agree because _____.
- I respectfully disagree with that idea because _____.
- I'm confused by _____.
- I was wondering _____.

Questioning

- Can you explain how _____?
- Can you explain why _____?
- I was wondering _____.
- I don't understand. Can you explain it to me in a different way?

2. Engage students in another number talk. After the number talk, ask the class to comment on what it noticed about how students used the various talking-to-learn moves and how these moves supported learning.

REFLECT

As our students become more skilled in talking about their mathematical thinking and listening and responding to the reasoning of others, their math learning deepens and spreads throughout the classroom community.

Use the following discussion questions to reflect with students:

- What are we learning about talking about math?
- How will this help build our powerful math community?

LET'S DO SOME MATH ACROSS OUR SCHOOL!

The experiences of solving and writing riddles support vocabulary development, logical reasoning, and a love of math. Riddles also provide opportunities for students to practice talking about mathematical ideas.

Share the riddle you wrote earlier with your students. Use the Talk Facilitation Moves in Table 7.1 to help students think through and solve the riddle together. Then, guide the class through the process of creating another riddle.

After you have written several riddles together as a class, students can write their own riddles with partners or independently (see Station Activity on p. 231). You can use the Color Tiles Riddles activity from Manipulatives and Models Matter or suggest that students write riddles about math content they are currently learning. Riddles can be written about any of these math topics:

- Whole number of any size
- Decimal numbers
- Fractional numbers
- Coins
- Angles, polygons, three-dimensional shapes
- Measurement tools

Post your students' riddles in the hallway outside your classroom so that others can admire your students' mathematical brilliance.

TEACHING MOVE: CONCEPTS FIRST, THEN VOCABULARY

Math vocabulary names mathematical ideas and relationships. When students understand and can use vocabulary related to the math they are learning, they can be more precise in their mathematical thinking and communication. Students need to understand a mathematical idea before they can meaningfully attach a name to that idea (Burns, 2022; Kobett & Karp, 2020). For instance, before students can attach real meaning to the word "polygon" they must see and touch a variety of polygons, noticing and talking about their attributes.

We want students to understand and use math vocabulary that names concepts related to the math content we teach (e.g., addend, sum, numerator, denominator), as well as words that describe the processes they engage in as mathematicians (justify, critique, represent, analyze).

REFLECT

- Identify one vocabulary word related to the math content you will teach this week. How can you provide students with experiences to develop their understanding of this mathematical idea before introducing the label for this idea?

- What is one vocabulary word related to a math habit that students will engage in this week? How can you support students in using this word to talk about their mathematical thinking?

CLASSROOM ROUTINE: REVISITING NORMS

Because of the winter break, January is an important time of year to revisit a classroom's norms. Chapin et al. (2022) told us that the norms of respectful discourse and equitable participation are essential to a powerful math community. Here are some examples of behaviors that indicate the presence of these norms:

WHOLE-CLASS DISCUSSIONS
• When explaining their math thinking, students automatically justify their ideas without prompting.
• Students listen attentively and respectfully to other students' ideas.
• Students use a polite tone of voice and respectful language when commenting on other students' ideas.

(Continued)

(Continued)

• Students ask questions and respectfully challenge other students' ideas.
• Students talk directly to each other
SMALL-GROUP COLLABORATION
• Students are mindful of the balance of voices in group conversations.
• Students encourage and support quieter group members in sharing their math thinking.
• Students check to make sure that all group members understand the task and the group's thinking about the task.
• When the group or a group member is stuck, the group works together to ask questions, share ideas, and use math tools to get unstuck.

Use these lists of behaviors to observe the math talk that takes place in whole-class discussions and small-group collaboration in your classroom.

Which of these norms are firmly in place? Compliment students on these important learning behaviors.

Choose one norm that could be strengthened. Talk with students about why this norm is important, and together develop a plan to grow this norm.

CLASSROOM ROUTINE: LEARNING TO PRESENT

When a student or student group shares their problem-solving work with the class, they have the opportunity to develop communication and presentation skills that will serve them well in future school and career contexts.

Take a few minutes to talk with students about what an effective presentation of mathematical work looks and sounds like. List ideas on an anchor chart or checklist that students can refer to as they prepare for a mathematical presentation and later as they reflect on their performance. This list might include:

- Display your work on a poster or with the document camera so everyone can see it.

- Explain your problem-solving process and solution clearly and slowly using a voice that is loud enough for everyone to hear.

- Ask the math community for questions and connections about the mathematical ideas.

- Ask the math community for feedback about your presentation.

During the presentation, community members should look at and actively listen to the student making the mathematical presentation. During the discussion of the work that follows the presentation, community members can use the talking-to-learn moves to support discourse and understanding.

CLASSROOM ROUTINE: HANDS-DOWN CONVERSATIONS

One way that teachers can support students' practice of essential discourse skills is by stepping out of the facilitation role, sometimes allowing students to collaboratively facilitate a math discussion on their own.

A hands-down conversation is a classroom routine developed by teachers Wedekind and Thompson to help their students learn the skills of listening and speaking in academically focused conversations. "Hands-Down Conversations are, most simply put, conversations that flow among students without the use of hand-raising, and in which the teacher is not the primary speaker" (Wedekind & Thompson, 2020, p. 3). In a hands-down conversation, students sit in a circle facing each other and the teacher sits outside the circle, listening to the conversation.

A hands-down conversation is guided by three simple rules:

- No hands. Listen for the space to slide your voice into the conversation.

- One voice at a time (more or less!)

- Listen closely to everyone's ideas. (p. 3)

(Continued)

(Continued)

You can easily incorporate hands-down conversations into your lessons with topics such as the following:

- A mathematics concept or topic (What do we know about subtraction?)

- A student's mathematical conjecture (What do we think about the strategy that _____ is suggesting we try?)

- An interesting mathematical idea from a student's math journal or exit ticket (Here's what a student wrote in their journal yesterday about decimal numbers: _____ What do we think about these ideas? What other ideas about decimals are important?)

As you listen to students talk during a hands-down conversation, you might take notes about:

- Patterns of participation

  - Who talks and who is not yet talking?

  - How do students participate in the conversation? Who shares strategies or conjectures? Who asks questions of other students? Who poses wonderings?

  - How are students listening to each other's ideas? How are they supporting each other in sharing mathematical ideas?

- Mathematical ideas discussed

  - What important mathematical ideas are students noticing or not yet noticing?

  - What misconceptions are present?

  - What mathematical connections are students making?

The data you gather during a hands-down conversation can support your instructional decisions and planning. You can also use this data to help the class reflect on and set goals for building their math habits and understandings.

STATION ACTIVITY: MATH RIDDLES AND JOKES

Provide examples of math riddles for students to solve and to use as models for writing their own riddles. Students can write riddles and leave them for classmates to solve, recording the answer on the back of the paper. You might occasionally choose a student-written riddle to share with the class for transition time or a warm-up activity.

You might also offer a selection of math jokes at this station. Here's a source to get you started: https://bit.ly/49J6Xx6 (Tapp, 2023)

Suggest tasks such as the following for using math jokes to support math learning and fun:

- Find a math joke that you think is funny and share it with someone else.

- Draw a picture to illustrate a math joke.

- Notice how math vocabulary is used in math jokes. Pick a math vocabulary word that your class knows and write a joke using that word.

- You might organize a Math Joke-a-Thon where interested students take turns telling their favorite math joke.

 Available as a downloadable resource on the companion website.

LITERATURE CONNECTIONS

***Usha and the Big Digger* by Knight** (2021) (Grades K–2)

Usha and her friends see a constellation of stars from different orientations and, as a result, describe it differently. In the end, they learn that our unique perspectives help them to understand ideas in different ways.

(Continued)

(Continued)

Activity: Use pictures of the Big Dipper and other constellations for number talks. How many stars do you see, and how do you see them? How do we all learn more by listening to each other's mathematical thinking?

***Marvelous Math: A Book of Poems* selected by Hopkins** (2001) (Grades 3–5)

This book of poems about mathematics captures emotions that are commonly associated with doing math and the relevance of math to real life. Individual poems can be used to prompt thinking and discussion about how we relate to mathematics. The book illustrates the value of talking about our mathematical experiences.

Activity: Write a poem to celebrate math or complain about math. Share your poem with someone else. How can poetry help you to think about what math is and how it is important in your life?

SPOTLIGHT ON BRAIN SCIENCE

Scientists have discovered a special kind of brain cell called a "mirror neuron." Brain scans of monkeys revealed that mirror neurons activate when a monkey picks up a peanut and when that same monkey watches someone else pick up a peanut. Mirror neurons send messages to the emotional system in our brains that allows us to feel empathy and to learn by watching others (Mirror Neurons, n.d.). These specialized neurons play an important role in building classroom norms and a sense of community.

You can learn more about mirror neurons in this short video: https://bit.ly/3VaUcXF

You can share the following information with students in a mini lesson or a station activity. You might also send this reproducible home along with the Family Newsletter to help families talk about these important ideas.

MIRROR NEURONS HELP US BUILD A POWERFUL MATH COMMUNITY.

Our brains have a special kind of brain cell called "mirror neurons." Mirror neurons begin to work when we watch another person do something. Mirror neurons make us feel excited when we watch our favorite sport. These neurons help us to imagine that we are playing the game. Our mirror neurons might cause us to automatically smile when we see someone smile or yawn when we see someone else yawn.

Image source: Istock.com/palau83

You can use mirror neurons to build a powerful math community in your classroom.

When you use fist bumps and cheers to show that you are proud of a classmate or your group for persevering with a challenging math problem, you activate your classmates' mirror neurons so they feel that same pride.

When you show curiosity about a math idea by noticing and wondering, you are sparking your classmates' mathematical curiosity.

And when you listen carefully to others share their math thinking, you are helping everyone in your class know that your classroom is a caring learning environment and that everyone's voice is important. You can use mirror neurons to build a strong classroom math community in which everyone thinks together and helps each other to grow as mathematicians every day.

How can knowing these scientific facts help you now and in the future?

Choose one of the following ways to think more about this important information:

1. Write down why these ideas are important to you and how you want to put them to work in your life.

2. Talk to a friend about these ideas. Decide together how you will put them to work.

3. Share these ideas with someone at home. Tell them why these ideas are important to you and how you plan to use them.

online resources Available as a downloadable resource on the companion website.

SPOTLIGHT ON EQUITY: SUPPORTING TALK

When we accept that talk promotes math learning, we are obligated to ensure that all students regularly talk about their mathematical thinking. The norm of equitable participation is complex, however, because it involves both human emotion and classroom culture. Chapin et al. explained:

> Standing up in front of your students and saying, "Everyone must talk," is unlikely to establish equitable participation in your classroom. Instead, equitable participation will depend on your instructional decisions – decisions that allow all students a chance to take the floor and that support students who have previously been reticent in class. (2022, p. 74)

Thoughtful facilitation of classroom talk can, over time, build a classroom culture of equitable participation. Here are three practices that support inclusive classroom dialogue:

- **Think time**—Provide generous amounts of thinking time before calling on students or asking volunteers to share.

- **Rehearsal**—Provide opportunities for students to practice talking to a partner or with a small group before asking them to share their thinking with the whole class.

- **Random selection of speakers**—At times, use randomization to call on students to share out. A variety of computer apps can be used to randomly select names. A popsicle stick can be randomly pulled from a can containing a set of sticks with student names. During collaborative group work, you might randomly preselect students to serve in the role of group reporter (e.g., the student whose first name is closest to M in the alphabet will share for the group).

Teachers need to think carefully about the needs of quieter students. Wedekind and Thompson (2020) suggested that we "shift away from the view of silence as purely negative and the silent child as a problem in need of fixing" and instead "seek to understand a child's silence" (p. 15). Students may be reluctant to talk for a variety of reasons, including:

- Students may be thinking and listening.

- Students may have internalized cultural norms that discourage talking.

- Students may feel that their ideas are not important or valued.

- Students may lack confidence in their ability to put their ideas into words.

Wedekind and Hermann Thompson (2020) reminded us: "As a classroom builds the dialogue community, it is crucial to recognize that each individual has different talk patterns and preferred indicators of listening. Cultural norms, family values, and individual hardwiring bring a beautiful diversity to human communication that we want to honor in our classrooms" (p. 62).

TRY IT

Think of a student who currently does little talking during whole-class discussions and small-group work. Observe this student over time to gain a sense of their communication habits and possible reasons for these habits. Then set aside a time to conference with this student. Share how talking about our math thinking supports learning. Invite the student to think with you about scaffolds that could make it easier to share their math thinking. Support the student in setting a small goal related to talking during math time. Together create a plan for specific things the student and you will each do to work toward achieving this goal. Schedule a time to meet again to monitor progress.

MATHEMATICAL ME: STUDENT JOURNAL AND PORTFOLIO

STUDENT JOURNALS:

Have students respond to the following questions:

- Why is it important to talk about my math thinking?

- How can I get better at talking about my math thinking?

Older students can write their responses in their Mathematical Me journals. Younger students can draw a picture showing their response, or you might gather responses during a class discussion or quick one-on-one interviews.

Review students' responses to monitor students' beliefs about mathematical communication. You might summarize this data by tallying or graphing the different ideas students offer.

Be sure to share and discuss this data with your class. Ask, "What does this data tell us?" and "What are some things we can do to continue growing as mathematicians?"

(Continued)

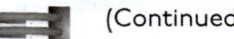

(Continued)

STUDENT PORTFOLIOS:

Have students record three to five math vocabulary words they've learned along with their definitions and an example of how each is used. On a sticky note, have students write the date and complete this sentence frame:

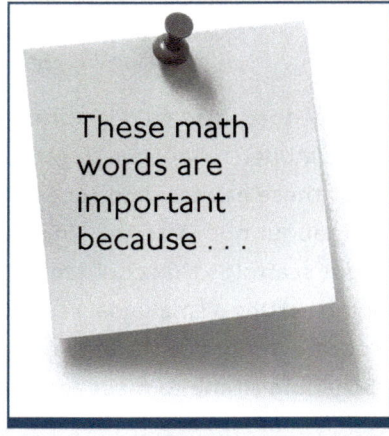

These math words are important because . . .

Source: Istock.com/Thammask-Chuenchom

FAMILY NEWSLETTER: WE TALK ABOUT MATH!

WHAT WE'RE LEARNING

This month we're learning about how talking about our mathematical thinking helps us to learn mathematics.

WHY IT'S IMPORTANT

When your child talks about math, they build math reasoning skills and strengthen their math vocabulary. Talking about math helps your child remember what they learn and feel more confident as a math learner.

Your child has learned about five talking-to-learn moves they can use to support their math learning:

- Explaining and justifying—sharing an idea and giving a reason or providing evidence

- Connecting—adding onto an idea or making a mathematical connection

- Conjecturing—suggesting a mathematical idea or strategy to try

- Critiquing—agreeing or disagreeing respectfully with an idea

- Questioning—asking a question of another student, the teacher, or the class

Your child has been practicing these moves using language stems like these:

Explaining and justifying

- I think _____ because _____.

- This makes sense to me because _____.

- This seems reasonable to me because _____.

- I chose this method because _____.

Connecting

- This idea connects with _____ because _____.

- This is similar to/different from _____ because _____.

- Another way to think about this is _____.

- Another strategy is _____.

Conjecturing

- I think we should try _____ because _____.

- I think the pattern is _____.

- I'm noticing _____ and wondering _____.

Critiquing

- I agree because _____.

- I respectfully disagree with that idea because _____.

- I'm confused by _____.

- I was wondering _____.

Questioning

- Can you explain how _____?

- Can you explain why _____?

- I was wondering _____.

- I don't understand. Can you explain it to me in a different way?

(Continued)

(Continued)

HOW YOU CAN HELP

Encourage your child to talk about their math thinking as they share about the math work they are doing in school. Also, encourage discussion of the mathematics you and your child do in daily activities outside of school.

The talking-to-learn moves are important learning and life skills in and out of school. You can help your child gain confidence talking about their thinking by modeling these moves yourself and encouraging their use in appropriate situations.

 Available as a downloadable resource on the companion website.

CHECKING IN ON OUR LEARNING: WE TALK ABOUT MATH!

- Why is it important to talk about my math thinking?
- How can I get better at talking about my math thinking?

This month's learning focus, talking about math, is integral to the other math habits. A class's mathematical discourse allows students to practice reasoning and sense-making and thus supports students' math learning. As teachers and students work to strengthen the class's mathematical discourse, the classroom community itself grows stronger and more supportive of the academic, social, and emotional needs of each community member.

MATHEMATICAL ME: EDUCATOR JOURNAL

- How has this month's learning focus supported your students' mathematical growth?

- How has this month's learning focus supported your growth as a math teacher?

- How has this month's learning focus supported your school's growth as a powerful math community?

LOOKING AHEAD

In February, we will look at the different ways that mathematical ideas can be represented. We'll consider the value of representing mathematics in multiple ways and the importance of seeing connections between different mathematical representations.

Questions for Reflection

- When you do math outside of school, how do you typically record and represent this mathematics?

- When do you find yourself sketching diagrams, tables, and other pictorial representations to think about and keep track of your mathematical thinking?

- What kinds of representations do you encourage students to use in their math learning?

Answers to Riddles at Start of Chapter

1) 10, 20, 40, 80

2) 2 pennies, 3 nickels, 3 quarters

3) The age must be 14, 42, or 70. We might reasonably say it has to be 42 because 14 and 70 are not likely ages for a teacher.

CHAPTER EIGHT

February

We Represent Math in Different Ways!

ESSENTIAL QUESTIONS

- What are the different ways I can represent math ideas?
- How does representing math in different ways help me learn?

THIS MONTH'S FOCUS

In a powerful math community, leaders, teachers, and students recognize that mathematicians represent their work in various ways. The representations that they create and analyze help them to deeply understand concepts.

Leaders and teachers know that doing and learning math has shifted from only one way, one method, and one answer to seeing the interconnectedness of strategies, models, and representations of answers. Instruction is focused on helping students see that their representations matter and are essential to understanding the math they learn.

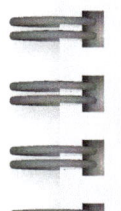

MATHEMATICAL ME: EDUCATOR JOURNAL

Take five minutes to respond to the essential questions. You will have a chance to reflect on these same questions and to have students respond to these questions at the end of the month.

ON YOUR OWN

LET'S DO SOME MATH! MATHEMATICAL TUG-OF-WAR

This wonderful problem is a classic from Marilyn Burns and engages you to think about complex ideas while remaining accessible by nature (Burns, 1982). As you solve the Mathematical Tug-of-War Problem, you will find that representing your thinking unlocks an understanding of the structure of the mathematics used.

There are three rounds in this tug-of-war, and it is your job to determine who wins. The contestants are Ivan, a specially trained dog; four acrobats; and five grandmas who have practiced for years.

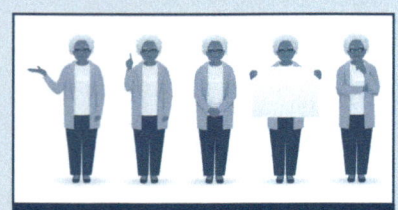  

Sources: Istock.com/Volhah; Istock.com/den0909; Istock.com/ONYXprj

The final round of the tug-or-war will be Ivan and three grandmas versus the four acrobats. Using the information from Rounds I and 2, can you figure out who wins?

ROUND I:

In Round I, with four acrobats against five grandmas, the results are even. Neither team can pull the other forward.

Sources: Istock.com/ONYXprj; Istock.com/Volhah

ROUND 2:

In Round 2, the results are the same. Ivan, on one side, and two grandmas and one acrobat are dead even.

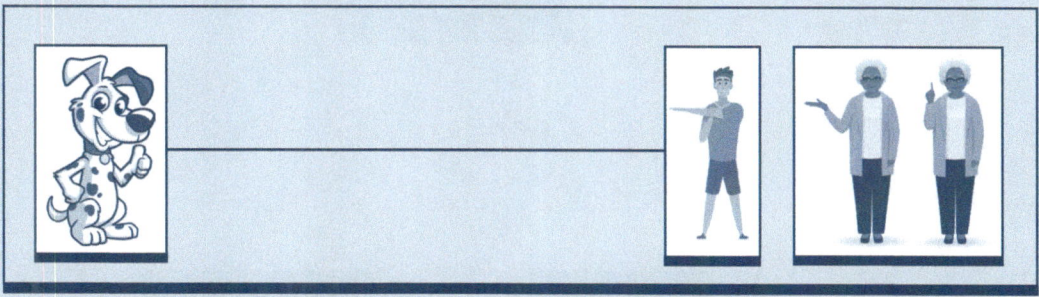

Sources: Istock.com/den0909; Istock.com/ONYXprj; Istock.com/Volhah

ROUND 3:

Round 3 is shown next. Ivan and three grandmas will be pulling against four acrobats. How can you represent your thinking and show who would win the final round?

Sources: Istock.com/Volhah; Istock.com/den0909; Istock.com/ONYXprj

Consider how you engaged in the habits of mathematically powerful people as you worked on the Mathematical Tug-of-War Problem. Save the representation(s) you used to reason about and solve the problem. You will share your representations with your colleagues during Together With Your Teaching Community.

 Available as a downloadable resource on the companion website.

Why This Focus

In the 1960s, American psychologist Jerome Bruner introduced the **CRA** or Concrete, Representational, Abstract approach to learning mathematics (Bruner, 1960). His work helped educators see that challenging math content can be learned at young ages as long as students have opportunities to grapple with ideas in multiple ways. Figure 8.1 demonstrates how each component of CRA is connected and how students can move back and forth between these approaches.

Figure 8.1 ◆ Concrete, Representational, Abstract (CRA) Approaches to Learning Math

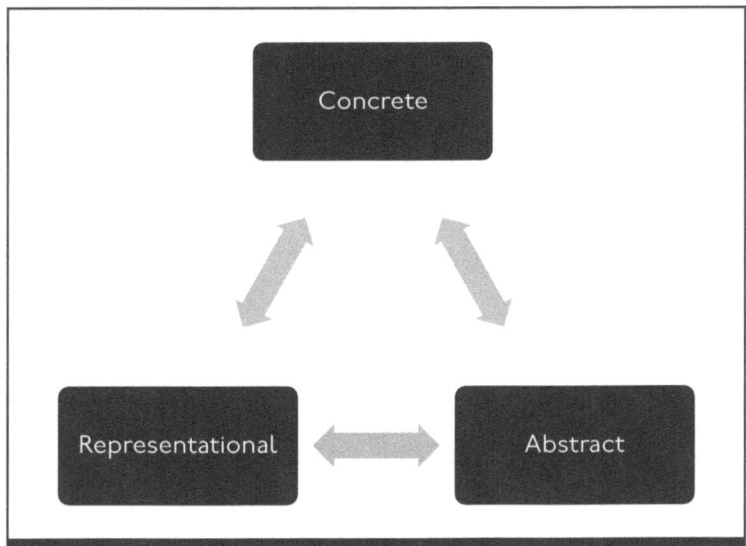

In their book *The Math Pact*, Karp et al. (2021) wrote, "Representations are commonly thought of in three different ways: (1) physical or concrete (e.g., hands-on), (2) semiconcrete or diagrammatic (e.g., pictorial, graphically represented), and (3) abstract or symbolic (e.g., written symbols- numbers, variables and operational symbols)" (p. 50). They added that:

"it is beneficial to use representations as tools because they

- Provide ways for students to examine and identify relationships

- Help students compare and contrast multiple depictions of a mathematical idea to determine generalizations or patterns

- Give students a way to communicate their thinking about a concept or process (Barrera & Santos, 2001), and

- Support deeper understanding of the mathematical content and practices than would otherwise happen if we only present these ideas with the use of mathematical symbols while hoping that students make accurate mental connections." (p. 50)

Concrete

Concrete experiences may include students using manipulatives to model, communicate, and make sense of mathematics. For example, when learning about fractions, we may offer students fraction bars to physically move around, put together, and compare fractions. In her book *About Teaching*

Mathematics, Burns (2015a) wrote, "Physical objects are important tools for teachers to use for teaching mathematics and for students to use for learning mathematics. . . . Manipulative materials have long been staples for math instruction, and for good reason" (p. 99).

Burns listed the benefits of incorporating manipulative materials into math instruction as follows:

- They help students make sense of abstract concepts.
- They provide students ways to test and verify ideas.
- They serve as useful tools for solving problems.
- They make learning mathematics more engaging and interesting.

Semi-Concrete Representational

Semi-concrete representations refer to "a pictorial representation of the concrete objects or a picture or graphic organizer that represents a problem situation or schema. They also include other types of representations such as number lines, graphs, diagrams, and tables" (Karp et al., 2021, pp. 50–51). For example, students learning about comparing fractions might draw a model of the fraction bars they have used, taking care to show the relationship between the pieces as they compare their sizes. They label the fraction bars and use the model to help them solve problems and communicate their findings.

Abstract

In mathematics, *abstract* refers to the equations, symbols, and variables used to communicate mathematically. Students are shown how to write symbols and equations so they can be precise in how they represent their mathematical thinking and solutions.

Students need to have opportunities to experience all three approaches of concrete, representational, and abstract so they have a deep understanding of mathematical concepts.

Representations in Five Categories

We can further break down representations to include real-life situations and spoken language as ways for students to access and make sense of math. Figure 8.2 shows the interconnectedness of the five representations.

Figure 8.2 • Connecting Representations

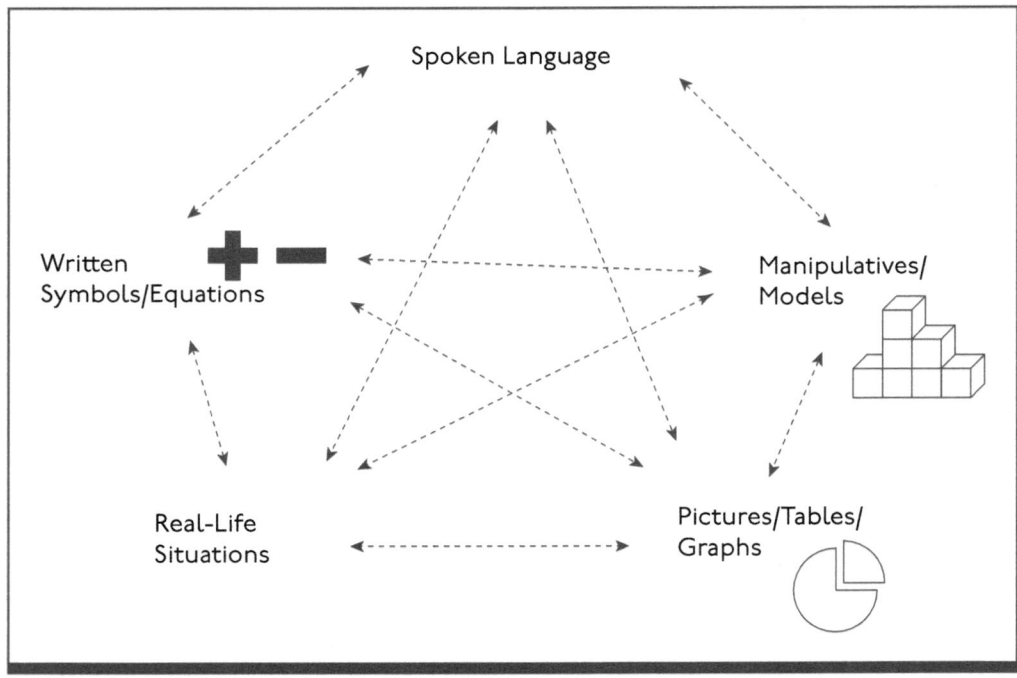

Source: Lesh, Post, & Behr (1987)

More information about the five representations is outlined as follows:

- **Written Symbols/Equations**: Record numbers, operation symbols, and variables.

- **Pictures/Tables/Graphs**: Illustrate ideas with drawn pictures, t-charts, tables, and graphs.

- **Spoken Language**: Speak about math with mathematical vocabulary to describe ideas.

- **Manipulatives/Models**: Use tools to think through concepts and create models.

- **Real-Life Situations**: Connect mathematics to real-life situations.

Each representation is interconnected. When math learners consider how a mathematical concept looks in more than one way, they are digging deeper into the content they are learning. For example, a student learning to multiply fractions may struggle to see why three times one fourth equals three fourths. However, a student who can use language to describe what happens when we multiply fractions may lead to other ideas of representation. A student might say, "Three times one fourth is the same as three groups of one fourth." A simple illustration demonstrating their thinking helps students visually see the operation and quantities at work.

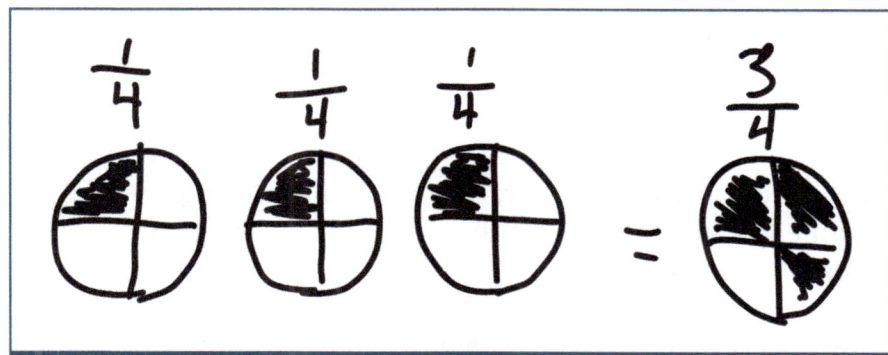

Learners also use multiple representations as a source for communication about strategies and solutions. If you recall the math habits learners use as they engage with mathematics, being able to justify your reasoning and precisely state your answer are essential. Math learners who represent their work in multiple ways, including concretely, semi-concretely, and abstractly, provide others with access to their thinking and ideas.

What's even more essential is that math instruction should center on the connections made between each of the different representations. Teachers support a deep understanding of mathematical content when they help students connect concepts to other concepts and representations. In Chapter 9—March, we will do a deep dive into the connections we help students make between the various representations.

Holly's Math Story:

As a student in the classroom, my math experiences were limited to abstract methods. The teacher told us how to solve a problem that was followed by time to practice problems just like it. I rarely had opportunities to see and feel the math and concretely experience it.

Gonzalez explored this idea when she wrote about how people tend to view mathematics:

> What is mathematics? What ideas, words, and images jump to mind when you try to define the subject? For many, the answer revolves around numbers, computation, and equations. Images conjured are often of a chalkboard covered in symbolic language or of one's own experiences in a mathematics classroom. (2023, p. 35)

Learners use multiple representations as a source for communication about strategies and solutions. Math learners who represent their work in multiple ways, including concretely, semi-concretely, and abstractly, provide others with access to their thinking and ideas.

The beginning of my teaching career looked very similar to my own experiences as a student. I taught students abstract methods such as algorithms for multiplying and dividing. As I had more experience teaching and spent time with colleagues, I recognized the need to incorporate concrete experiences, specifically using math manipulatives for students to make sense. It took a long time and many mistakes to comfortably use manipulatives for students to experience math. Although incorporating concrete experiences, including using manipulative materials, is certainly a step in the right direction, I could tell many students still needed time to transition from the concrete materials to abstract written equations and symbols.

In addition to concrete experiences, students need opportunities to consider how their hands-on work is translated into a visual representation. We can think about moving from physical objects to written methods and mathematical equations as fluid, and students are moving between CRA as they reason and make sense. In Chapter 9—March, we will further discuss the importance of making connections between the representations so students understand the mathematical content deeply.

Students' math learning is enhanced when they can move concepts between concrete representations, pictorial or graphical representations, and then abstract representations. The remainder of this chapter is dedicated to showing the power of using a variety of representations for students to uncover the mathematics they are learning.

Tapping Into Your Experience

As we build our powerful math community, we need to reflect on our own experiences with using representations. Take a moment to consider your math teaching and learning. Check off which representations you have used.

- **Written symbols and equations**: Using numbers, operations, and variables to write expressions and equations

- **Pictures, tables, and graphs**: Drawing pictures that represent mathematical ideas, creating a table of values, or creating various graphs

- **Spoken language**: Speaking about math concepts, strategies, and solutions

- **Manipulatives and models**: Using manipulatives such as connecting cubes, base-ten blocks, or tiles and making a model

- **Real-life situations**: Seeing the mathematics in everyday life

Consider:

- What types of mathematical representations did you use to solve the Mathematical Tug-of-War Problem?

- Why do you think you chose to use these representations?

- How can you push yourself to use some representations that aren't as familiar to you?

- Where can you use different representations in your teaching?

- Where might you use different representations in your life outside of school?

- How might strengthening your use of representations be helpful as a math learner and teacher?

Why Multiple Representations?

Asking students to represent their thinking in a variety of ways is one way to help them have a deep conceptual understanding of mathematical content. Boaler shared that:

> In 2013 research scientist Joonkoo Park and Elizabeth Brannon reported on a study in which they found that different areas of the brain were involved when people worked with symbols, such as numbers, than when they worked with visual and spatial information, such as an array of dots. The researchers also found that mathematics learning and performance were optimized when these two areas of the brain were communicating with each other. We can learn mathematical ideas through numbers, but we can also learn them through words, visuals, models, algorithms, tables, and graphs; from moving and touching; and from other representations. But when we learn by using two or more of these means and the different areas of the brain responsible for each communicate with each other, the learning experience is maximized. This has not been known until recently and has rarely been made use of in education." (2019, pp. 103–104)

"While one representation is beneficial, students will develop a richer, more detailed view when they see the same concept through multiple perspectives."

—SanGiovanni et al. (2022, p. 157)

How might knowing about multiple representations and their effect on students' performance influence your instruction?

Encouraging Students' Use of Multiple Representations

SanGiovanni et al. (2020) wrote, "Challenging students to go beyond finding a single answer and to create multiple representations opens up the possibilities to discuss and explore concepts and make conceptual connections between concepts and even procedures" (p. 86).

We can use questions to facilitate student thinking beyond a single representation. Questions we might ask include:

- How might you draw a picture that represents your thinking?
- Is there a math tool that might help you think through the problem?
- What is an equation that would represent your picture?
- When might you see this math in the real world?
- What would be some words to describe what you have modeled?

Connections to the Math Habits

The math habits discussed throughout this book are at the heart of learning math. Math learners use these habits to make sense, think through problems, and communicate their reasoning, among many other things. The representations discussed in this chapter are interwoven through the math habits.

Representation: Spoken Language

- Math learners are encouraged to justify and critique the reasoning of others. When students speak about their work, strategies, or solutions, they are representing their work through language.

Representation: Pictures, Tables, Graphs

- Math learners model their ideas. When students use or create a model, they are representing their work pictorially.

Representation: Manipulatives

- Math learners use tools that support their thinking through problems and communicating to others their ideas. When students use concrete objects, they are representing their work physically.

Representation: Equations

- Math learners use numbers and symbols in math. When students write equations, they are representing their work abstractly.

TOGETHER WITH YOUR TEACHING COMMUNITY

Since We Met Last (10 minutes)

In January, you and your colleagues gathered classroom data about your students' mathematical communication.

You might use the following protocol to share and reflect on this data with your math teaching community:

- Share your data with a colleague. (5 minutes)

- As a group, discuss what you learned from the data and how these insights can help you support your students' growth as mathematicians. (5 minutes)

Let's Do Some Math Together! (10 minutes)

In small groups, share your solutions to the Mathematical Tug-of-War Problem. Does your group agree on the solution? If not, explain and justify your solutions to each other. Take time to share the similarities and differences in the representations that were used to solve the problem.

Discuss:

- What math habits did you and your colleagues engage in as you justified your solutions to the Mathematical Tug-of-War Problem?

- How were the representations that were used similar and/or different from one another?

- How does each representation model the problem?

Building Our Expertise (30 minutes)

When we see the importance of representations for student learning in mathematics, we are more aware and intentional of the learning experiences we set up for students. In his article "Representation—Show Me the Math," former National Council of Teachers of Mathematics (NCTM) president Fennell (2006) stated that teachers should be purposeful in planning lessons that incorporate the models, tools, and representations that students will use for deepening their understanding of concepts. We must consider the mathematics being taught, work through the problems we ask students to solve, and then consider what might students use to represent the math problems. By engaging in this practice, we have a better idea about what we may see from students, which allows us to prepare questions and scaffolds along the way (https://bit.ly/3VaUcXF).

Together with your colleagues, select a math lesson or task students will be engaging in. Use the following steps to consider what representations you will include in the learning experience:

1. Do the math task/problem that students will solve.

2. List an example of each type of representation that goes with the problem or task.

3. Determine which representations you will use. Consider which representation(s) make sense for the problem and will deepen students' understanding of the content.

4. Consider what questions and discussions you will have with students to make connections between the representations.

Questions to connect representations:

• How does your model/drawing connect to your equation?

• How are these two models alike? Different? How do they both represent the problem?

• How do the parts of your equation connect back to the problem?

How might collaborating with your colleagues to preplan the representations you will use shift your teaching of mathematics?

How could you set up a regular time with your math community to develop opportunities for incorporating representations into math lessons?

Let's Try It (10 minutes)

Intentionally selecting and using representations for teaching and learning math rests on several factors:

1. The representations students are currently using and comfortable with.

2. The mathematical content and how it lends itself to certain representations.

3. The representations that would help students to make connections within concepts.

With your teaching community, select one of the data collection options below. Bring the classroom data to next month's meeting to discuss with your colleagues.

Record this task and deadline in your calendar or planner.

OPTION 1: WHAT REPRESENTATIONS ARE STUDENTS USING?

Select a problem for students to solve that can be represented in a variety of ways. Ask students to solve the problem in any way that makes sense to them. Collect the student work to analyze with a colleague at next month's meeting. Consider sorting the work into piles by type of representation. For example, place work together that shows only numbers/symbols, work that shows pictures or drawings, instances where students used concrete objects/manipulatives, and those that verbally shared their solutions.

OPTION 2: TALLYING THE REPRESENTATIONS USED IN YOUR MATERIALS

In the next month, tally the different types of representations your curricular materials are using. You may create a list of the different representations, including Written Symbols/Equations, Spoken Language, Pictures/Tables/Graphs, Real-Life Situations, and Manipulatives/Models. With every lesson, make a tally for each representation you see and the representations students are asked to use. At next month's meeting, analyze the results with a colleague. What do you notice?

OPTION 3: CLASSROOM DISCUSSION

Show students the image of representations shown below. Ask students to think about which representations are familiar to them and which feel more unfamiliar. Ask them to give an example of each type of representation. Make some notes about the discussion that you can debrief with a colleague at the end of the month.

(Continued)

(Continued)

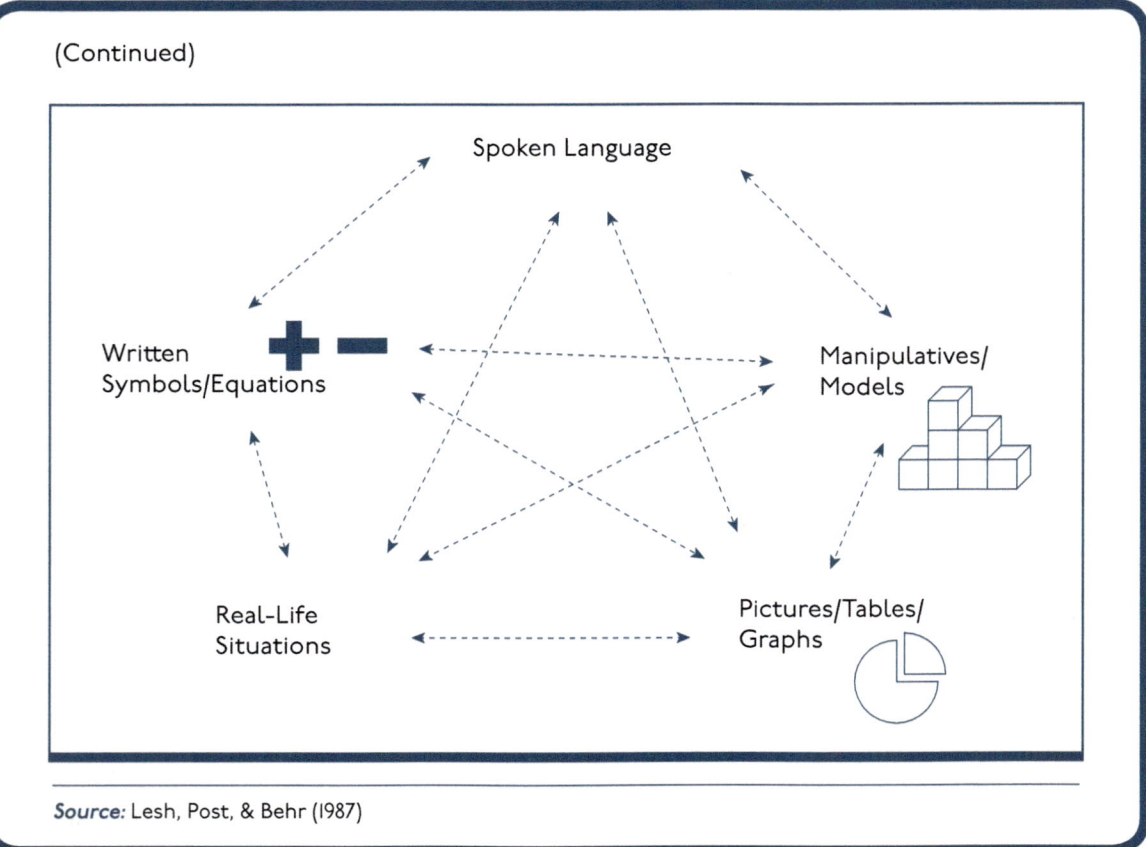

Source: Lesh, Post, & Behr (1987)

ADDITIONAL PROFESSIONAL LEARNING ACTIVITIES

This month's additional professional learning experiences engage educators in the content of fractions. According to Parrish and Dominick (2016), authors of *Number Talks*, "proficiency with fractions is underdeveloped; our own memories of learning about fractions consists primarily of dividing pizzas, pies, and candy bars and memorizing procedures for computation." (p. 2) This section aims to offer your math teaching community other models, tools, and representations of fractions that help students to make sense of fractions and fraction computation.

Math Talks: Fractions Talks

This month's math talk is a Fraction Number Talk. Use the same number talk facilitation protocol as in November:

- The teacher is a facilitator and recorder (refrain from teaching or determining correctness)

- All solutions that are shared are recorded, and students are asked to defend their answers (this is an opportunity for students to determine correctness and find and fix their mistakes)

Together with your teaching community, give the problem below a try.

You might ask teachers to name a fraction that represents a relationship in Figure 8.3. Ask teachers to justify their thinking, explaining how their fraction is represented in the image.

Figure 8.3 • Grocery Store Math

Discuss the experience:

- How might this experience help students to see mathematics in the real world?

- Below are some other mathematical representations. How could you use one of the other representations to describe the mathematics in Figure 8.3?

  ○ Spoken language

  ○ Manipulatives and models

  ○ Equations and symbols

  ○ Pictorial and graphical

- What math habits did you engage in during the fraction talk?

Manipulatives and Models Matter: Cuisenaire Rods

Cuisenaire Rods are ten multicolored rods that go from the smallest of I centimeter in length to the largest at 10 centimeters in length. They are incredibly versatile and can be used to count, add, subtract, multiply, divide, model fractional relationships, operate on fractions, and more. An illustration of Cuisenaire Rods is shown below.

| orange |
| blue |
| brown |
| black |
| dark green |
| yellow |
| purple |
| light green |
| red |
| white |

Burns (2015a) offered suggestions to teachers when using manipulatives in the classroom, which include free exploration. She wrote, "Giving time for free exploration is essential. Free exploration allows students to satisfy their curiosity" (p. 103). It's also a great time to internalize the structure and imagine how the tool can be used to help you learn math.

Together with your team, take out the Cuisenaire rods and explore them in any way that makes sense to you. You might create designs, line them up, talk about their relationship to one another, or see where the manipulation takes you. You may even consider exploring the rods and creating different representations such as pictures, equations, words, or even real-world examples.

Next, focus on the relationship between the rods. How does each color relate to the next? What mathematical vocabulary can you use to describe the relationships?

REFLECT

- What relationships did you find between the different rods?

- How did exploring this tool help you think about using them for instruction with students?

Game Time

This month's game offers a meaningful experience for players to add fractions using a math tool. The experience in Manipulatives and Models Matter is a helpful prerequisite to playing this game.

FRACTION TRAINS VERSION 1

Materials:

2 players

Cuisenaire rods (brown, purple, red, white)

Fraction spinner

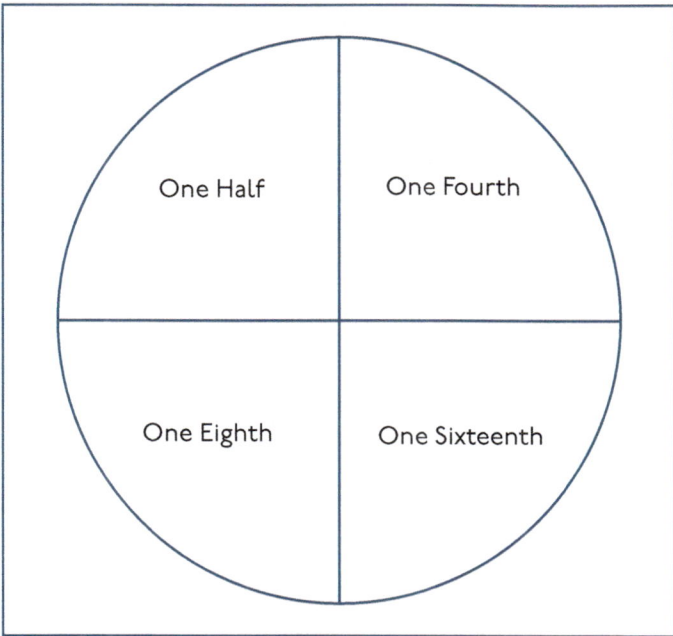

(Continued)

(Continued)

Paper clip

Pencil

*Tip: To make the spinner work, place a paperclip on the spinner with a pencil holding it in place in the center. With one hand holding the pencil, use your other hand to flick the paperclip so it spins around the pencil.

Challenge: Be the first player to cover two brown rods.

Note: For this game, players should use two brown rods to represent the whole. As often found in the exploration of Cuisenaire rods, the orange rod typically serves as the "whole." However, a wonderful feature of Cuisenaire rods is that they aren't labeled and allow for more flexibility in their use. Since this game works best with two brown rods serving as the whole, take time to look at the relationship of the two brown rods to other color rods. Before playing the game, discuss the fractional name for each of the rods used in the game. For example, since two brown rods are a whole, one brown rod is one-half.

Game directions:

1. You and your opponent should place two brown rods end to end in front of you.

2. Decide who will go first.

3. On your turn, start by spinning the spinner.

4. The result tells you which size rod to place under the whole.

For example: Player 1 spins one-fourth

5. Your opponent repeats Steps 3 and 4.

6. Take turns until one of you makes a train the same length as the whole. You are the winner.

7. If you or your opponent spin a fraction which makes your train longer than the whole, you lose your turn.

BONUS! FRACTION TRAINS VERSION 2

Game directions:

1. Place two brown rods end to end in front of you.

2. In this version, spin and remove the fraction that the spinner lands on.

3. On your turn, you have a choice:

   a. Spin the spinner and remove the fraction.

   b. Replace your brown pieces with equivalent pieces. For example, exchange one brown rod for two purple. *You may not exchange pieces and spin on the same turn.

   c. Spin the spinner and do nothing.

4. You may only remove pieces you have or equivalent pieces to what was on the spinner.

5. The first player to remove their entire train wins.

Discuss the experience:

- How does this game help students make sense of fractional representations?

- What other fraction concepts does this game address?

- What math habits did you engage with while playing this game?

 Available as a downloadable resource on the companion website.

IN YOUR CLASSROOM

Classroom Story

Mrs. Bahri has a lively group of second graders who love starting their day with dot talks sharing their flexible thinking. It's apparent that students have ample opportunity to practice number sense and math is enjoyable for them. Although her class is certainly demonstrating a powerful math community, Mrs. Bahri wanted to explore how to support her students with solving challenging word problems. Students could solve addition problems in which they joined like objects such as;

- *Ari had 5 roses and cut 4 more roses to add to a vase. How many roses were now in the vase?*

Now, students would be solving problems that involved combining different types of objects within a category. For example:

- *Aaliyah picked 5 daisies. She then picked 3 wildflowers. How many flowers does she have?*

Mrs. Bahri considered what teaching moves might support her students in working through this challenging concept without taking away the rigor. She often uses think–pair–share during morning number sense routines, so she wondered about focusing on the language students use prior to solving and working out problems on paper. Mrs. Bahri also wondered what type of drawing or representation would support the students as they spoke about their problems and solutions.

She decided on a model in which students could see a grouping of items with a label that fit all objects in the group (Figure 8.4).

Figure 8.4 • Category Diagram

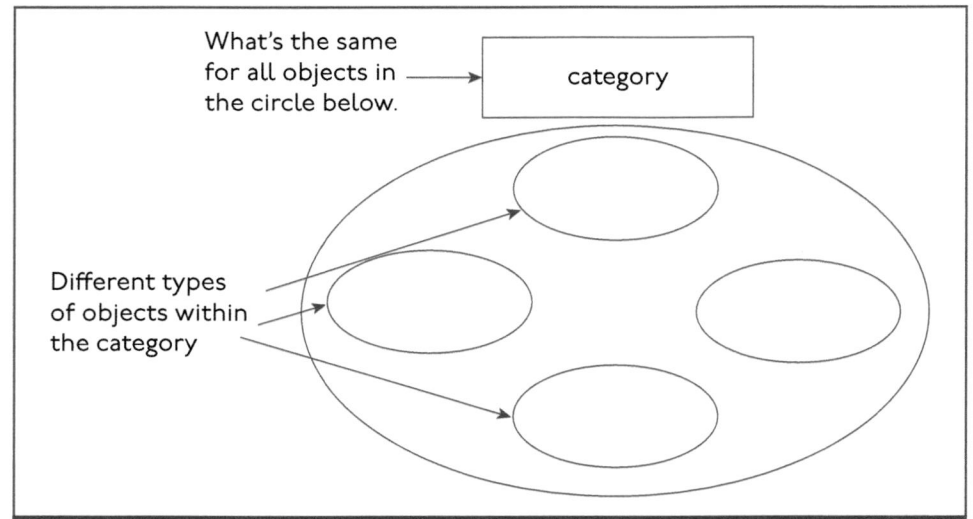

Mrs. Bahri created some high-interest and relevant topics for second graders. These would serve as the categories for each group. Her goal was to engage students in the context of the problem, use language to discuss their ideas orally, and then connect the experience to equations that represented the problem.

Mrs. Bahri started her lesson with all students together using flowers as the category for her class example. Students were asked to turn and talk about what types of flowers they knew of and she would call on four different students to provide examples in the category of flowers.

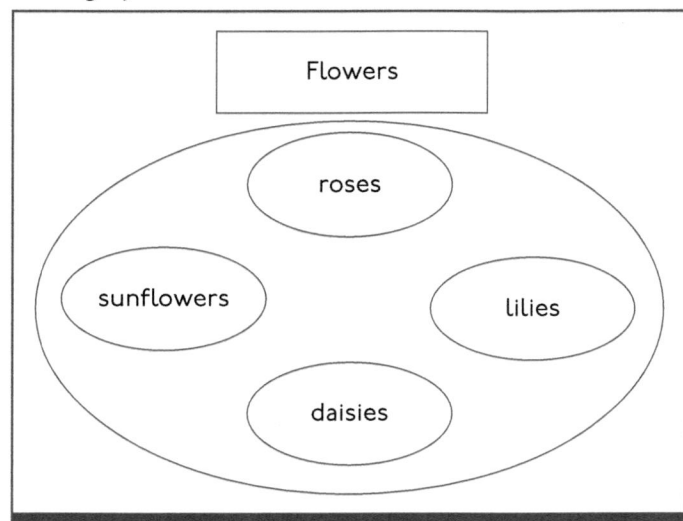

Mrs. Bahri then asked students to talk to a partner about a question they could ask for which the answer would be "How many flowers?" Khaleesi and Dajana share with each other.

Khaleesi: (pointing to the ovals) I had five sunflowers and five daisies. How many daisies did I have?

Dajana: Do you mean how many flowers?

Khaleesi: Oh yeah. How many flowers do I have?

Dajana: Ten

Khaleesi: Yes. Your turn.

Dajana: I have seven roses and two roses. How many flowers do I have?

Mrs. Bahri called the students back together and asked for a few students to share what they and their partner discussed. As students talked through their conversations, some still had a hard time moving into two different objects within one category. Many students shared adding roses to roses or lilies to lilies, but very few were able to share different flowers within the same category. Mrs. Bahri brought the students back to the question of "How many flowers?" if you picked from two categories. This approach seemed to support them, and as she had a few more students share, she then recorded the equations with labels that represented the language students were using.

Five sunflowers plus five daisies equals ten flowers.

$5 + 5 = 10$ flowers

Students then set off to work in small groups of four or five. They were tasked with listing different objects that fit into the category they were given. One group had the category of "drinks" and used the model to fill in with their ideas.

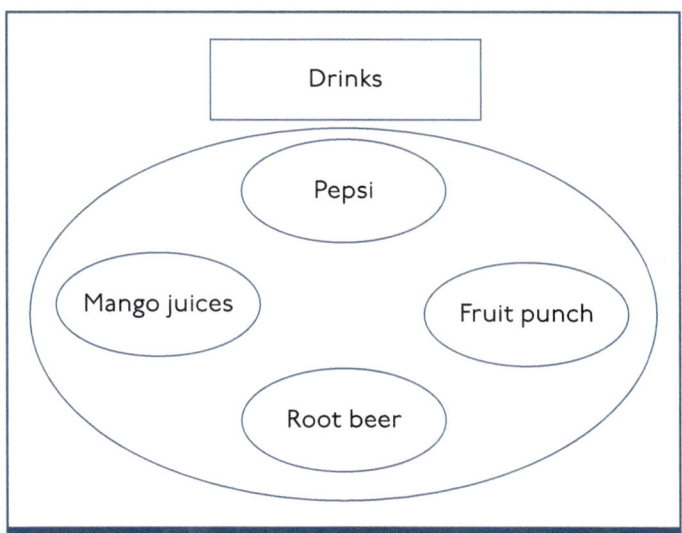

The four girls were eager to share. They began by going around the circle and orally sharing a word problem that would answer "How many drinks in all?" By the end of the experience, students were getting the hang of naming each addend as a different part of the category and then answering with the sum as the title of the category.

After the experience, Mrs. Bahri reflected on the lesson and the use of the model and spoken language to support students. She felt students were quite successful in being able to create equations for word problems within categories. Mrs. Bahri thought that having students pick their own context and then sharing orally with a partner allowed them to practice before committing to an equation. Spoken language was a support that allowed students to process their ideas verbally before committing them to paper. She also felt the visual model served as a support in helping students represent the problem. Students could use the representation and language, which would lead to writing equations. By representing the math problem in various ways, students were able to make sense of the task at hand.

Anchor Lesson

Representing Math in More Than One Way

WHAT?

In this lesson, students will generate multiple representations for a single math problem. They will communicate their reasoning and solutions to others and reflect on the importance of representing their ideas in more than one way.

YOU NEED:

Chart paper

Markers

Math task that can be represented in a variety of ways

Paper

Manipulatives

HOW?

1. Write "Using multiple representations" on the top of an anchor chart.

2. Engage students in a math task that lends itself to multiple representations. This includes problems in which students can draw a picture, use manipulatives, make a table or graph, and use language and equations to represent quantities and ideas. You might find a task from your curriculum, or you can use one of the problems listed below:

   • Kindergarten and first grade: Half of the people in a family wear glasses. How many people could be in the family, and what would a drawing of the family look like?

   • Second and third grade: One fourth of a class brings their lunch to school each day. How many students could be in the class, and how many bring lunch?

   • Fourth and fifth grade: Two or more fractions are added together to make one half. What could those fractions be?

3. Give students a sheet of paper folded into three sections. Students should write the different representations that they used to solve and model the problem at the top of each column. For younger students, fold the paper into two sections and support them in determining what representations they could use to solve the problem.

   • For example: Tell students on one side, they should draw a picture to represent their thinking, and on the other side, they could complete an equation that represents their drawing.

   • _____ wear glasses + _____ don't wear glasses = _____ people in the family.

4. Make a variety of tools available to students.

5. Circulate as students work and prompt them by asking, "What could you use to represent the problem so you can think about it differently?"

6. Once students have had the opportunity to solve the problem, bring the whole group together to reflect on the experience.

7. Ask students what representations they can use when they are working on math problems. Under "Using multiple representations," record student responses. You may even write an example next to each representation.

(Continued)

(Continued)

REFLECT

Our students gain a deeper understanding of content when they generate multiple representations and notice the connections between them. We can support this habit by expecting students to show their work in a variety of ways and discuss their representations with the classroom community.

Use the following discussion questions to reflect with students:

- How are the different representations similar? Different?

- Why might it be important to represent a problem in more than one way?

LET'S DO SOME MATH ACROSS OUR SCHOOL!

Students begin to have experiences with algebra as early as kindergarten. Finding an unknown is a key concept that students will experience each year in mathematics classrooms.

The Mathematical Tug-of-War Problem invites students to use representations to reason about and solve the problem. Ivan, the acrobats, and grandmas can be represented with concrete manipulatives such as color tiles to help students think through the problem. Older students can also represent the problem symbolically.

GRADES K-3

Use a simplified version of the Mathematical Tug-of-War Problem. For example, one grandma is equal to two acrobats, and two acrobats and one grandma are equal to Ivan. What would happen in the final round of Tug-of-War?

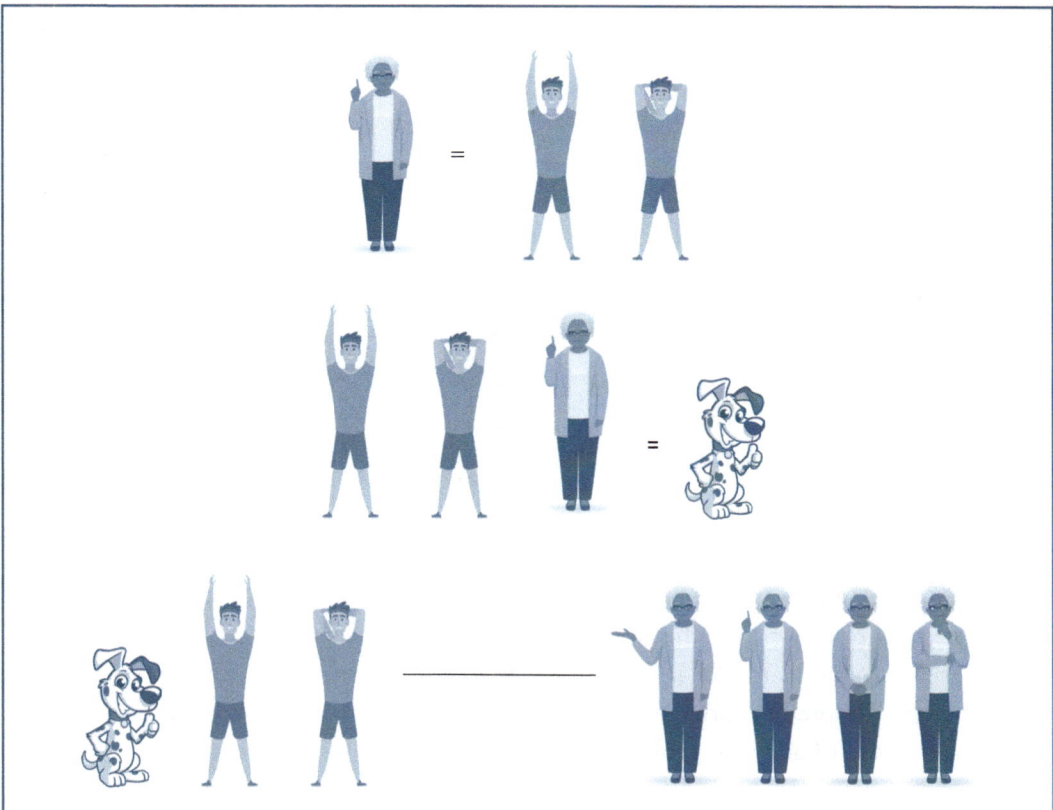

Image sources: Dog by Istock.com/Volhah; Old lady by Istock.com/den0909; Young man by Istock.com/ONYXprj

(Continued)

(Continued)

Encourage discussion between students using questions like these:

- If one granny went against one acrobat, who would win?

- Who do you think is the strongest of all the players?

GRADES 4-5

Use the original Mathematical Tug-of-War Problem with older students. As they grapple with the problem, encourage discussion between students using questions like these:

- How did you represent the different players in the tug-of-war?

- If one granny went against one acrobat, who would win?

- Who do you think is the strongest of all the players? How might that information help you solve the problem?

- If you had to represent each player with a quantity, what would you use? Why?

Post your students' work outside the classroom where others can admire and be inspired by your student mathematicians.

TEACHING MOVE: REPRESENTATION ANCHOR CHART

Throughout this chapter, a variety of representations are referenced. For students to know about and use multiple representations to support them in solving problems, they need time to experience them and reflect with their peers. Think back to how Mrs. Bahri focused on a model and oral language for students before writing equations. She selected a couple of representations and gave students time to explore them so they could make sense of addition problems.

Create an anchor chart with students that introduces them to representations that are appropriate for their grade level. Use a sample problem to demonstrate what the representation looks like, how it's defined, and other times that it could be used in math.

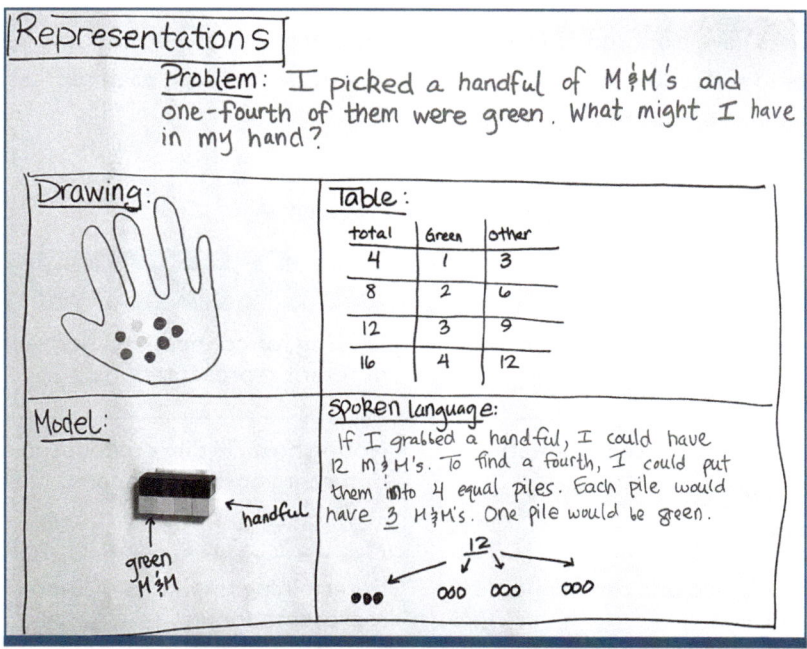

Representations

Problem: I picked a handful of M&M's and one-fourth of them were green. What might I have in my hand?

Drawing:

Table:

total	Green	other
4	1	3
8	2	6
12	3	9
16	4	12

Model:

green M&M
handful

Spoken language:
If I grabbed a handful, I could have 12 M&M's. To find a fourth, I could put them into 4 equal piles. Each pile would have 3 M&M's. One pile would be green.

REFLECT

- How will spending time reviewing the types of representations support students in using them while solving problems?

TEACHING MOVE: CONNECTING QUESTIONS

In Chapter 10—April, we will discuss the importance of helping students make connections. We must recognize that asking students to represent their thinking in multiple ways requires us to also help students see why those representations all reflect the same concept and that connecting each representation to another is how we deepen our understanding. We don't require students to represent their thinking in multiple ways simply to create more work for them and ourselves, but the interconnectedness of models, drawings, numbers, equations, and spoken language paint a broader picture of mathematical ideas.

(Continued)

(Continued)

Once students have explored different representations, we can use connecting questions for students to reflect on the mathematics that is shown through each representation. Sample questions are listed in Table 8.1.

Table 8.1 • Using Connecting Questions to Support Students

CONNECTING QUESTION	HOW IT SUPPORTS STUDENTS
How is _____'s representation similar to _____'s representation?	Looking for connections between two different representations
How is _____'s representation different from _____'s representation?	Noting how certain content looks different yet still connected
Why might you use one representation over another?	Determining the limits of each representation
How do the multiple representations help you understand _____? (the content)	Seeing the benefit of creating multiple representations for understanding mathematical concepts

APPLY

While planning for an upcoming math lesson, consider what representations students may use. Make a list of the representations and select a question from the table above that you will ask at the end of the lesson. Think about and write down some responses you expect to hear from students.

REFLECT

- How can connecting questions help students see the interconnectedness between representations?

- How did preplanning a question focus students on connecting their representations?

See Chapter 9—March for more ideas on Connecting Representations.

CLASSROOM ROUTINE: WHAT'S THE REPRESENTATION?

This routine will help students build the habit of using representations to help make sense of problems before attempting to solve them. It also strengthens students' fluency in using multiple representations.

Here are some steps to get you started with What's the Representation.

1. Prepare a challenging problem for students from your curriculum resources or choose a problem below.

   - K: How could two friends share four cookies?

   - 1: How could two friends share 12 cookies?

   - 2–3: How could three friends share 12 cookies? Four friends? Six friends?

   - 4–5: How could four friends share five cookies?

2. Read the problem together with students. Invite students to discuss what the problem is about and what it is asking them to figure out.

3. Review the representations that you introduced in the anchor lesson. Ask, "What representations might you use to help you think through the problem?"

4. Use a think–pair–share strategy to help students generate and discuss ideas. Ask students to think independently, then share with a partner, and finally have the whole class share ideas for different ways of representing the problem. Encourage students to consider concrete, semi-concrete, and abstract representations that might help them to make sense of and solve the problem.

5. Allow time for students to work on the problem as you circulate and encourage students to try out different representations.

6. Debrief by bringing the class back together for a discussion. Ask questions such as:

   - What different representations did you try?

   - How were your representations helpful to you in solving the problem?

   - Why can one problem be represented in so many ways?

   - Which representations felt easier to create? More challenging?

   - When would (type of representation) make sense to use?

STATION ACTIVITY: REPRESENTATION CREATOR

Students need practice with creating representations for a variety of problems. Teachers must make time to talk about the representations and their connection to problems and quantities. The station activity for this month will allow students to practice creating representations for different math problems.

PREPARING THE STATION:

1. Prepare two sets of cards. One set should include the name of representations students know and have used. The other set should have numbers, expressions, or equations.

   - In the illustration that follows, cards with words indicate the different representations students are familiar with. The other cards should have numbers, expressions, or equations that are grade level appropriate. Each example includes sample student work.

2. Make space for students to display their work. Ideas include a bulletin board, chart paper, open counter space, or even hallway walls.

IMPLEMENTING THE STATION:

1. Students will select one of each card. They will create a representation of the number, expression, or equation that was drawn. For example:

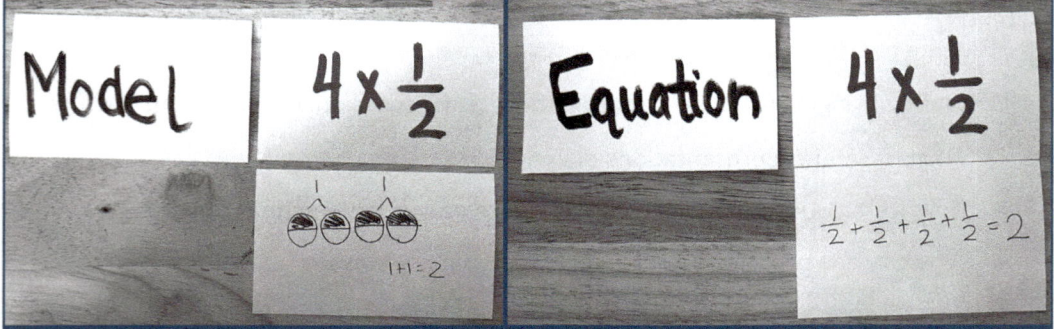

2. Students will display their representations in the designated area for others to view and for the teacher to use once students have had a chance to visit the station.

3. Once students have had time to interact with the station, reflect with the class using the discussion questions listed below.

For younger students, consider offering students a worksheet with limited choices. Students would select one from each column and show their work.

CREATE A REPRESENTATION	
CIRCLE ONE:	**CIRCLE ONE:**
Draw a picture	*Source:* Istock.com/Tetiana Holovanova *Source:* Istock.com/sabelskaya *Source:* Istock.com/mariaflaya
Write an equation	
Use a math tool	
Explain with written words	
Show your work:	

(Continued)

(Continued)

Discussion questions:

- What are some similarities you notice among the representations?

- What are some differences you notice?

- Which representations do you find easier to create? More challenging?

 Available as a downloadable resource on the companion website.

LITERATURE CONNECTIONS

If You Were a Plus Sign **by Shaskan** (2009) (Grades K–2)

This book discusses the mathematical convention of using a plus sign in equations to represent the mathematical operation of addition. Mathematics vocabulary is introduced, and addition stories are represented with pictures, numbers, and words.

Activity: Besides the plus sign, what other mathematical symbols do you know and use? Make a poster to show others how to use symbols to represent mathematical ideas.

Maryam's Magic: The Story of Mathematician Maryam Mirzakhani **by Reid** (2021) (Grades 3–5)

Maryam Mirzakhani was a famous Iranian American mathematician. She saw and represented mathematical ideas in art and stories. Her struggles and successes demonstrates multiple math habits.

Activity: Maryam Mirzakhani loved stories and she loved art. She used both of these passions to represent important mathematical ideas. How can you use your personal passions to think about and represent mathematical ideas?

SPOTLIGHT ON BRAIN SCIENCE

Our brain uses different regions to store different types of knowledge. When we represent a mathematical idea in different ways, it helps our brain construct multifaceted understanding of that idea. This idea can then be readily retrieved and applied in new contexts because it can be accessed from multiple regions of the brain (Willis & Willis, 2020).

The process of representing mathematics in multiple ways can be thought of as "brain glue" (p. 135). Concrete and semi-concrete representations are not just needed by young children. Math learners of all ages need to engage all of their senses as they explore and make sense of mathematical ideas and problems. We can do this by making it a habit to represent math in multiple ways.

You can share this information with students in a mini-lesson or a station activity. You might also send this reproducible home along with the Family Newsletter to help families talk about these important ideas.

REPRESENTING MATH IDEAS IN MULTIPLE WAYS MAKES OUR BRAINS STRONG.

When you represent a math idea or problem in different ways, your brain holds these memories in different storage areas (see Figure 8.5).

- When you move manipulatives, your brain stores this memory in your frontal lobe or your cerebellum.

- When you draw a diagram, a table, or a graph, your brain stores this memory in your parietal lobe.

- When you look at different math representations, your brain stores this memory in your occipital lobe.

- When you discuss different ways of solving a problem, your brain stores this memory in your temporal lobe.

(Continued)

(Continued)

Figure 8.5 • Where Memories Are Stored in the Brain

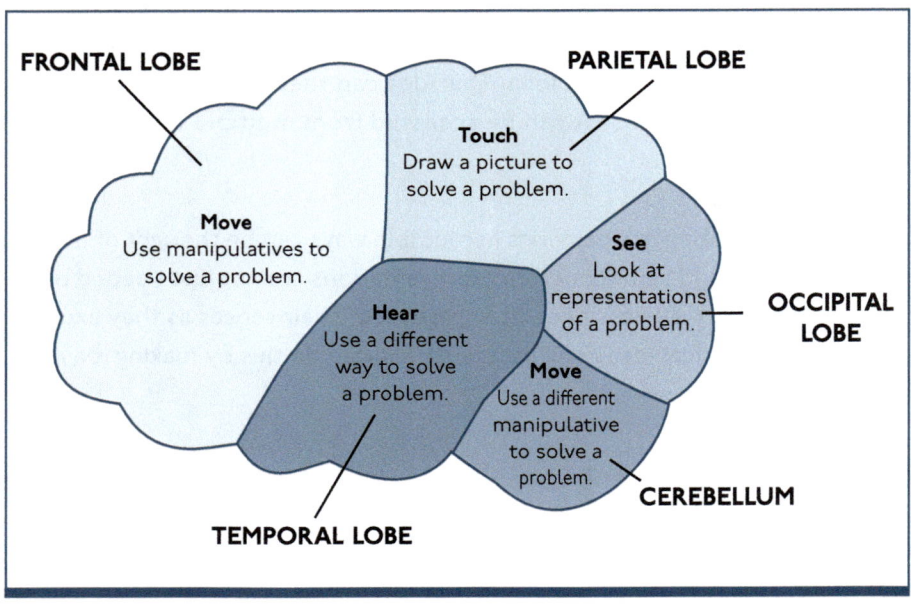

Image source: istock.com/Yulia

Because your memories are stored in multiple places in your brain, your understanding of the problem or the idea is stronger. It is easier for you to use this idea or strategy in the future because you can find it in different places in your brain. And because these different areas of your brain communicate with each other about this idea or strategy, the pathways between these regions become stronger and faster.

So, when you represent math in multiple ways, you make your brain stronger.

Choose one of the following ways to think more about this important information:

1. Write down why these ideas are important to you and how you want to put them to work in your life.

2. Talk to a friend about these ideas. Decide together how you will put them to work.

3. Share these ideas with someone at home. Tell them why these ideas are important to you and how you plan to use them.

online resources ▸ Available as a downloadable resource on the companion website.

SPOTLIGHT ON EQUITY: REPRESENTATIONS THAT VALUE DIVERSITY

"Incorporating history into the teaching of mathematics gives life to the work being taught"

—(Gonzalez, 2023, p. 161)

Representation of number is often limited to our understanding of the base-ten number system that we use. But mathematics is everywhere and all around us having historical roots in diverse groups of people. Gonzalez (2023) stated, "Every single civilization of which we know has a numeration system." She then went on to share how many of the systems we likely learned about had their roots in European civilizations. However, we miss opportunities to share about other non-European civilizations that have influenced other numerical systems such as the Babylonians. Their number system was based on the power of 60, leading to what we know as our measures of time (pp. 158–159). By recognizing, talking about, and spending time with mathematical contributions from other civilizations, we expand our mathematical view.

TRY IT

Show students the table of ancient number systems (Figure 8.6).

Figure 8.6 • Ancient Number Systems

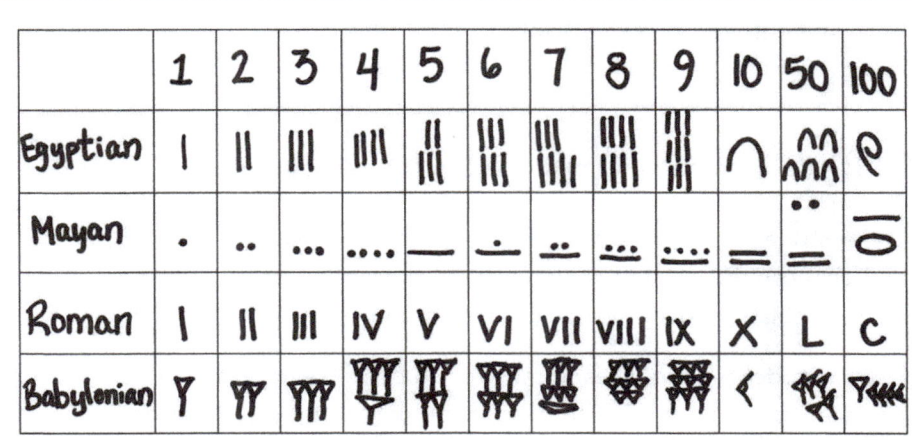

online resources — Available as a downloadable resource on the companion website.

(Continued)

(Continued)

Questions to facilitate discussion:

- What do you notice?

- What do you wonder?

- What connections can you make to what you know about the number system we use?

Looking for similarities and differences from our number system to these ancient number systems can broaden our students' appreciation of history and their worldview of mathematics.

MATHEMATICAL ME: STUDENT JOURNAL AND PORTFOLIO

STUDENT JOURNALS:

Have students respond to the following questions:

- What are different ways I can represent math ideas?

- How does representing math in different ways help me learn?

Older students can write their responses in their Mathematical Me journals. Younger students can draw a picture showing their response, or you might gather responses during a class discussion or quick one-on-one interviews.

Review students' responses to monitor students' beliefs about mathematical representation. You might summarize this data by tallying or graphing the different ideas students offer.

Be sure to share and discuss this data with your class. Ask, "What does this data tell us?" and "What are some things we can do to continue growing as mathematicians?"

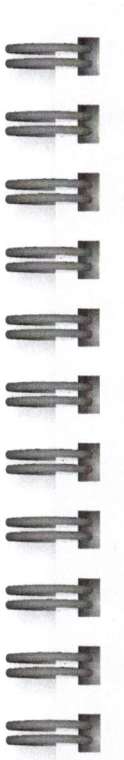

STUDENT PORTFOLIOS:

Have students choose a piece of work that shows a mathematical representation they are proud of. On a sticky note, have them write the date and complete this sentence frame:

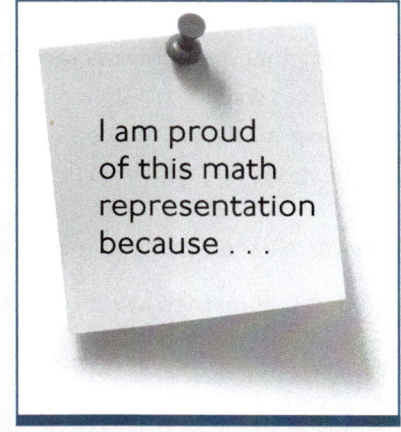

I am proud of this math representation because . . .

Source: Istock.com/Thammask-Chuenchom

FAMILY NEWSLETTER: WE REPRESENT MATH IN DIFFERENT WAYS!

WHAT WE'RE LEARNING

This month we are learning about how mathematicians represent their thinking in a variety of ways.

WHY IT'S IMPORTANT

Research has shown that people learn at a much deeper level when they see and engage in content in many different ways. In mathematics, students can use physical objects, drawings, pictures, tables, graphs, spoken and written language, numbers, variables,

(Continued)

(Continued)

equations, and more to think about math concepts and to communicate their reasoning and understanding. The more opportunities that children have to think about math with visual representations of mathematical ideas, as well as language, the more likely they will have a solid understanding of mathematical concepts.

HOW YOU CAN HELP

This month, your child is learning that mathematics is represented in the real world. As you go about daily activities such as walking or driving to school, shopping in the grocery store, or taking a walk in your neighborhood, notice some mathematical representations around you. Here are some questions you might ask:

- Where do you see numbers? What do those numbers mean?

- What images do you see that feel mathematical to you? What do they represent?

- Are you seeing any graphs or charts? Take a moment to look at them and ask, "What does this graph or chart show?"

- Where do you see math in nature? What mathematical ideas are represented?

 Available as a downloadable resource on the companion website.

CHECKING IN ON OUR LEARNING: WE CAN REPRESENT MATH!

- What are different ways we can represent math ideas?
- How does representing math in different ways help me to learn?

This month's learning focus, creating and using representations, is essential for learners to deeply understand mathematics. Our brains make strong connections when content is represented in a variety of ways and when we take time to notice the similarities and differences between these representations. Students need ample opportunities to explore a variety of representations and to think about how they are connected to math concepts.

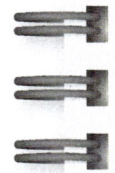

MATHEMATICAL ME: EDUCATOR JOURNAL

- How has this month's learning focus supported your students' mathematical growth?

- How has this month's learning focus supported your growth as a math teacher?

- How has this month's learning focus supported your school's growth as a powerful mathematical community?

LOOKING AHEAD

In March, we will explore how mathematicians make connections between concepts and representations and how math is connected to their lives.

Questions for Reflection

- How are math concepts and skills in your grade-level curriculum connected?

- How do these concepts and skills connect to the mathematics your students learned in earlier grades?

- How do you see mathematics connected to your life personally and professionally?

CHAPTER NINE

March
We Make Math Connections!

ESSENTIAL QUESTIONS

- How does making math connections help me learn?
- How can math connections help me solve new problems?

THIS MONTH'S FOCUS

In a powerful math community, leaders, teachers, and students make strong connections between mathematical concepts, mathematical representations, and from mathematics in the world to themselves. They know that math isn't just something we do inside the four walls of school and only on paper. The "doing" of mathematics is everywhere and all around us, permeating the things we do, what we say, and how we think.

Educators help students to notice mathematical connections, and they model the habit of making connections for students. They know that making connections supports memory and is therefore an essential learning strategy.

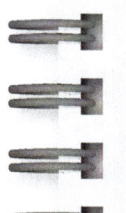

MATHEMATICAL ME: EDUCATOR JOURNAL

Take five minutes to respond to the essential questions. You will have a chance to reflect on these same questions and to have students respond to these questions at the end of the month.

ON YOUR OWN

LET'S DO SOME MATH!
MY FAVORITE REFRESHMENT

One of the sweetest sounds in my (Holly's) day is the "click" and "fizz" of opening a carbonated drink! I look forward to the time of day I can reach in the fridge, open a can, and take that first spicy sip of sparkling water. When I clean up my cans (yes, multiple), I often wonder how much sparkling water I drink in a week. Month? Year?

Image source: Istock.com/spaxiax

Take a moment to consider your go-to drink. How much of your favorite drink do you think you drink in a year? Devise a way to determine the number of cups and gallons of your favorite drink you have in a year.

For example, if I have two cans a day, with each can being one and one-half cups, I drink a total of three cups of sparkling water a day.

3 cups × 365 days a year is 1,095 cups a year

Save your work and the solution you came up with to share with your colleagues. Consider how you engaged with the habits of mathematically powerful people as you thought about and solved this problem.

 Available as a downloadable resource on the companion website.

Why This Focus

Holly's Math Story:

Sitting on my deck one summer evening, I was staring at the giant windowless wall that made up our western view. My husband and I wanted to be able to entertain more, and we wondered if we could build a large frame with white canvas covering the area to make a large outdoor screen. We would get a projector and could watch movies outside on our deck like a drive-in (or walk-in!) movie. It didn't take long to put together the project, and when it was done, we were excited to share it with our friends and family. I wanted to tell people the size of the screen, so I asked my husband how large he had made it.

The first issue was that I hadn't realized TV screens aren't measured by height or width but rather by their diagonal. The second was how I would get the measurement of the diagonal now that the screen was in place and out of reach.

I had a genuine moment of excitement when I realized that this would be the first time I would put the Pythagorean Theorem to use in my life! I had the length and width dimensions, which I could then use to calculate the diagonal.

I share this story because it's fascinating to hear how many people believe that math is no longer connected to them now that they aren't students in school. When I travel, people ask what I do, and when I share that I am a math teacher, they often tell me their horror stories of what happened to them in school and are often relieved that they no longer have to do math. But isn't math everywhere? Don't we use it every day?

In her book *Becoming the Math Teacher You Wish You Had,* Zager (2017) shared a story about asking her mom "what is math?" Although her mother had always used math in her adult life in running businesses and everyday experiences such as balancing checkbooks, she stated, "Math is when they hand you a sheet of paper, and it has a word problem you don't understand on it" (p. 2). Zager wrote that many people have such a narrow view of what it means to do and experience math. This response is too common among students and adults.

How do we move beyond the notion that math is only in school and help people see the connections of mathematics that are everywhere?

Helping students see math's **connections** to their lives is essential for student math identities, curiosity, and deep understanding and love for mathematics. As educators help to build students' personal connections to mathematics, they also build connections from concept to concept, connections from one representation to another, and connections to other content areas.

Connections Between Concepts

Tapper (2012) stated that connecting important information to other key concepts is essential. Students can make strong connections by comparing concepts, organizing ideas, reflecting, and discussing their thinking (p. 136). Educators guide learners through these connections using the Concrete, Representational, Abstract (CRA) approach discussed in Chapter 8—February. Figure 9.1 shows the CRA cycle and the interconnectedness of concrete experiences, connected representations, and abstract ideas.

> "When students perceive mathematics as a rich web of interconnected ideas, as opposed to a collection of disconnected facts and rules, they have the underpinnings to develop their confidence and agency in mathematics."
>
> —Huinker and Bill (2017, p. 121)

Figure 9.1 • CRA Approach to Learning Mathematics

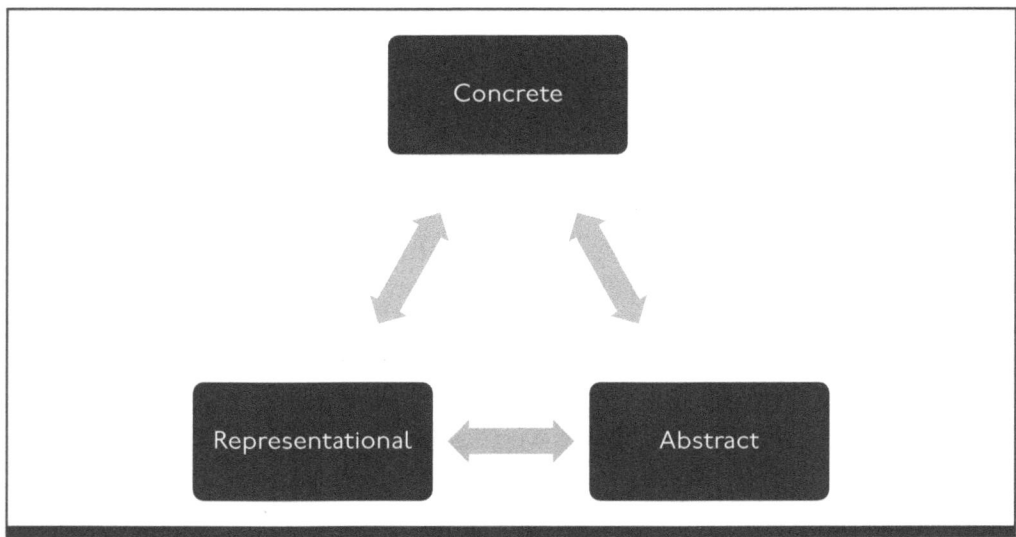

Let's take the process of making connections between concepts through the CRA cycle. When teaching third-grade students to multiply, we can connect to what students know about addition. We will use the following sample problem to demonstrate this idea:

> Each person in your group of four needs five toothpicks for our science experiment. How could we figure out how many toothpicks your group will need to take?

Students may use actual toothpicks or a math manipulative, such as connecting cubes, to model the process of each person getting five toothpicks. Students concretely represent the problem to help them think about how to quantify the total. The teacher then may ask students to consider how they could find the total number of toothpicks from the model. As students share, they may say that they could count by fives a total of four times (5, 10, 15, 20), or they could add five to itself four times (5 + 5 + 5 + 5). The teacher records these strategies on the board as an abstract model, or equation, that quantifies and represents the problem. As students work on similar problems, the teacher may introduce a representational model, such as a drawn picture, that represents the problems with circles representing each quantity in the problem. For example:

> For a new investigation, students will be in groups of three. Each person needs three toothpicks for the experiment.

The teacher may move to a representational model to show how to think about and quantify the problem.

The teacher would record the addition equation, 3 + 3 + 3 = 9, alongside the multiplication equation, 3 × 3 = 9. To demonstrate the connection from addition to multiplication, the teacher might say, "When we have repeated addition such as three plus three plus three, we can write this as an addition equation or we can write this as multiplication. Multiplication is equal groups of, and we have three equal groups of three. Therefore, we can also write this mathematically as three times three."

The purposeful movement between concrete, representational, and abstract mathematical representations ensures that students continue building strong, interconnected neural pathways.

Students' understanding of concepts and their connection to other concepts need to be intentionally considered when educators plan and prepare lessons. The purposeful movement between concrete, representational, and abstract mathematical representations ensures that students continue building strong, interconnected neural pathways.

Connections Between Representations

In their book *Talk Moves*, Chapin et al. (2022) wrote, "When students are able to connect ideas and concepts to procedures and representations, learning is especially robust. Thus, one of the things we want to talk about in mathematics class is how concepts and relationships among concepts are connected and how these ideas connect to what students already know" (p. 102). We can refer back to the five different types of representations that were introduced in Chapter 8—February. Figure 9.2 shows the five representations.

Figure 9.2 ◆ Connecting Representations

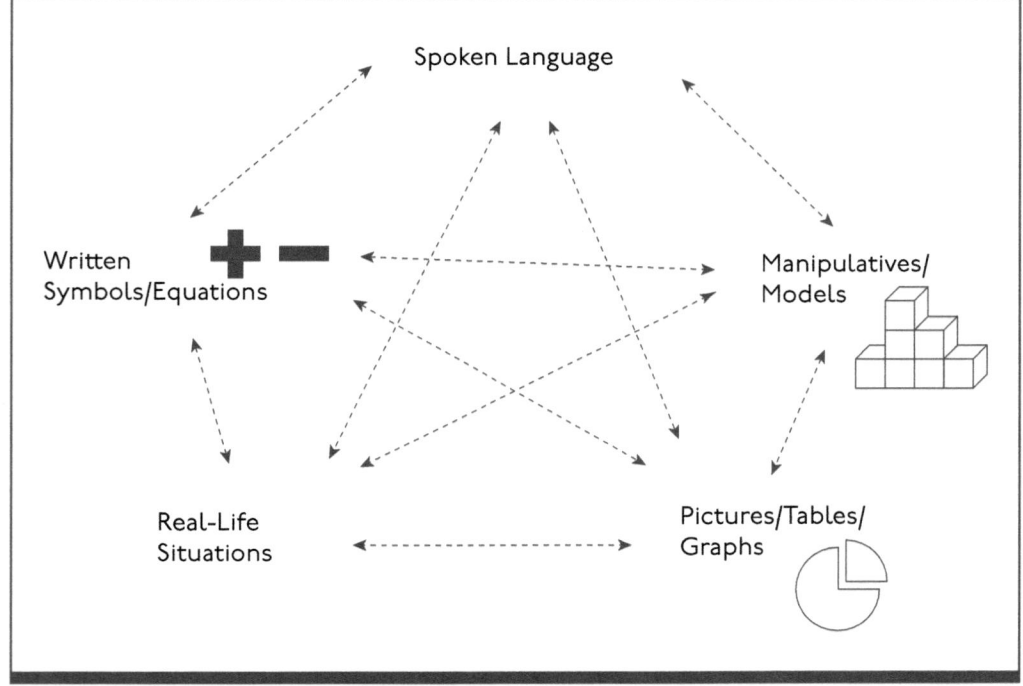

Source: Lesh, Post, & Behr (1987)

Each representation is useful on its own to help students make sense of math. However, students can better retain information and learn and speak about concepts at a deeper level when these representations are used together and explicitly connected. The goal is for students to have **representational fluency**, the ability to flexibly use and translate between different mathematical representations. Imagine a student is learning about the concept of multiplication. We may start with a real-life situation:

A bakery window displays their muffins, as shown following. How could you find out if there are enough muffins for your class of 15 students?

Image source: Istock.com/LightFieldStudios

We might start by asking students to think about and share their ideas for finding out if there were enough muffins. Students could use language to describe the picture and scenario to help them make sense of the problem. For example, they might say that there are four groups of three, which is equal to 12, so there wouldn't be enough to feed 15 people. Students are taking a real-life situation and describing the mathematics with spoken language.

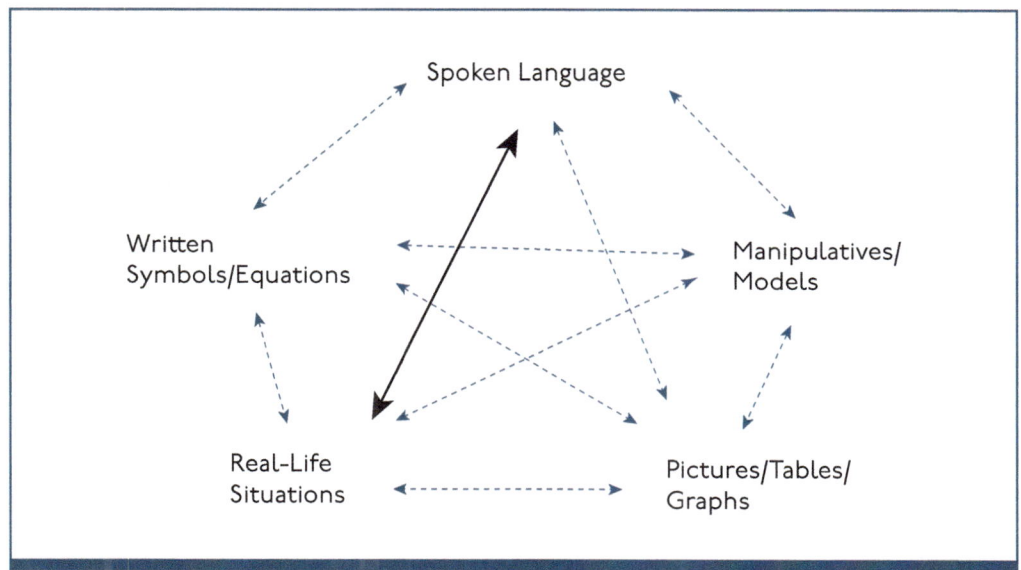

Source: Lesh, Post, & Behr (1987)

Next, we might ask students to think about how to write an equation to represent the number of muffins in the display case. Using their prior knowledge, students may say 4 + 4 + 4 or 3 + 3 + 3 + 3 to represent the rows and columns. Having used language to describe what they see, they can put quantities to their words to write equations.

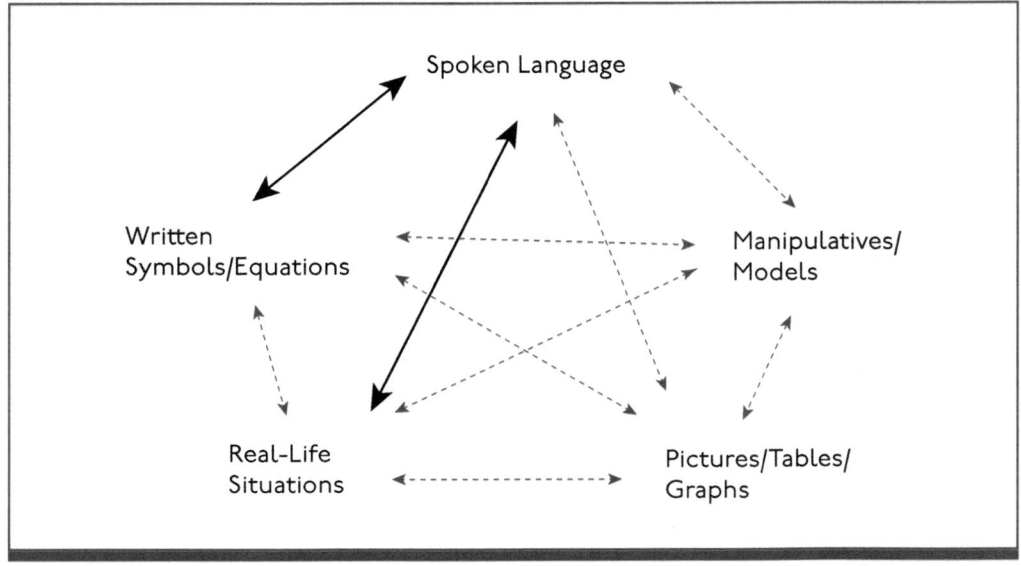

Source: Lesh, Post, & Behr (1987)

Students use their language, background knowledge, and a real-world scenario to quantify the situation. During this experience, teachers may move students back and forth between written symbols and spoken language as they introduce a new way to record equations with a repeated addend. They may say, "three rows of four can be written as 4 + 4 + 4, or it can be written as 3 × 4. This means three groups of four. Where do you see the three groups of four in the bakery picture?"

The teacher's intentional connection of three types of representations helps students better understand the concept. Boaler (2015, p. 184) shared, "Curriculum standards often work against connection making, as they present mathematics as a list of disconnected topics. But teachers can and should restore the connections by always talking about and valuing them and asking students to think about and discuss connections."

Think about a curriculum standard you are currently teaching. What are some ways you can help your students make connections to this new concept to support and deepen their learning

In this chapter, we will explore the connections that math learners make to themselves, between mathematical concepts, among different representations, and to other content.

Tapping Into Your Experience

Consider the role mathematics plays in your life. How do you see yourself connected to mathematics as you move about your day? It's helpful to think about personal experiences outside the school walls and your connections with mathematics in relation to learning in whatever role you encompass.

Create a two-column chart with the left column labeled "Connections to math in my life" and the right column labeled "Connections to math within my professional setting." Take five minutes to write down words or phrases that describe how math is connected to your personal life and how math may be connected to you in a professional setting.

Table 9.1

Connections to math in my life	Connections to math within my professional setting

After creating and reviewing your list, consider the following questions:

- Were there areas within your life that you hadn't thought about before? Did it surprise you?

- How might reflecting on your list help you in your role within your math community?

Making Connections Using the Math Habits

The habits of mathematically powerful people are designed to reflect mathematics within the math learner. Each habit represents a mathematical connection that you make to yourself.

1. I expect math to make sense
   - Mathematicians ask themselves what *connections* they have to the problem as they strive to make sense.

2. I love challenging math problems
   - Mathematicians *connect* doing math to solving challenging problems and feel a sense of accomplishment from that work.

3. I learn from math mistakes
   - Mathematicians *connect* doing math to a series of mistakes that eventually lead them to answers.

4. I see myself as a mathematician
   - Mathematicians see math as *connected* to themselves and can identify the mathness within them.

5. I talk about math
   - Mathematicians use communication to *connect* their ideas to others.

6. I represent math in different ways
   - Mathematicians use representations to *connect* context to quantity and from concrete to abstract.

7. I make math connections
   - Mathematicians recognize the essential skill of *connecting* ideas, representations, and content. They see *connections* between mathematics and the real world.

8. I look for and use patterns
   - Mathematicians use patterns to *connect* prior knowledge to new learning and discover new ideas.

Using Classroom Conversation to Make Connections

Student learning deepens when students have opportunities to talk about their thinking and reasoning. Chapin et al. (2022) stated, "As students talk about their reasoning and what relationships they grasp, they externalize their understanding. This helps them to build individual mental connections regarding concepts" (p. 94).

- When has there been a time that talking about your thinking has helped you develop clarity and understanding?

- How are the conversations occurring in your classroom helping students make connections in their math learning?

- What other math talk opportunities might you structure to support connection-making?

The Effect of Making Connections on Memory

When we connect new learning to existing understandings, this new learning is more easily remembered and can be more easily retrieved and used. Kobett and Karp (2020, p. 53) wrote, "A memory that has more connections to other memories (aka prior learning) is infinitely easier to retrieve because the student can access the memory from multiple connection points in a mental network of ideas." The more connections we make from existing understandings to new learning, the better it is remembered and the more easily it can be retrieved. When we support and expect students to continually connect their new learning to things they already know and can do, they end up with a strong network of connected understandings that can be easily accessed and applied in future learning and problem-solving work. On the other hand, when we teach skills and concepts in isolation, we make it difficult for students to connect this new learning to existing knowledge.

Notice the two images below. The one on the left demonstrates a single isolated idea, and if it's cut (forgotten), there are no other connections back. However, when we make connections to ideas and concepts we already know, we have a neural pathway that can be accessed through many routes, as demonstrated in the second image.

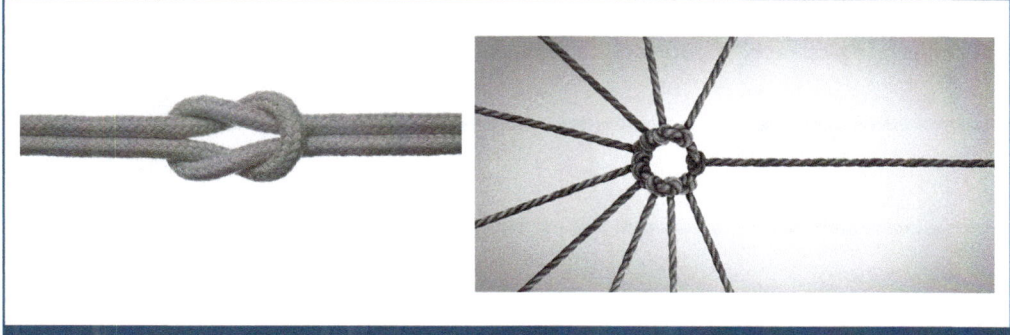

Image sources: Istock.com/grandaded; Istock.com/wildpixel

> When we support and expect students to continually connect their new learning to things they already know and can do, they end up with a strong network of connected understandings that can be easily accessed and applied in future learning and problem-solving work.

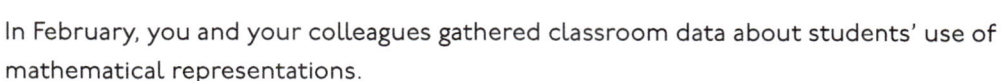

TOGETHER WITH YOUR TEACHING COMMUNITY

Since We Met Last (10 minutes)

In February, you and your colleagues gathered classroom data about students' use of mathematical representations.

You might use the following protocol to share and reflect on this data with your math teaching community:

- Share your data with a colleague. (5 minutes)

- As a group, discuss what you learned from the data and how these insights can help you support your students' growth as mathematicians. (5 minutes)

Let's Do Some Math Together! (10 minutes)

In small groups, share your number of cups and/or gallons consumed in a year of your favorite drink. Discuss the different ways your group determined the volumes.

Discuss:

- What is something in the world that would compare with that same volume?

- What math habits did you and your colleagues use when solving the problem?

- What connections were you able to make to other similar volumes in the world?

- How are the strategies each of your colleagues used connected to each other?

Building Our Expertise (30 minutes)

Part of the power in making connections between concepts and representations is that students begin to see the connectedness of mathematics from year to year. Former National Council of Teachers of Mathematics (NCTM) President Gojak stated:

> Although it is important to think about the connections among concepts within the grade level or courses that we teach, it is also important to reflect on the connections across grade levels. This work involves thoughtful discussions with other colleagues about the way that concepts are taught and the potential linkages among those ideas. Many of us learned mathematics as isolated pieces of information. Taking a mathematical concept and considering how it originates, extends, and connects with other concepts across the grades will help teachers to develop a deeper understanding. It is then that we can plan instruction that ensures that our students regularly make connections to help them make sense of the mathematics they are learning. (2013, para. 6)

With your colleagues, select a math topic that spans K–5. Suggestions include addition and subtraction, place value, geometrical shapes, or equations. Use the following steps to start a conversation about the vertical connections between that concept:

1. On your own, write down a couple of big ideas that students are learning around the chosen concept at your grade level. You may even choose to include an example.

2. If applicable, check in with a colleague who teaches the same grade level. Discuss your big ideas.

3. Starting with kindergarten and going up or starting with fifth grade and moving down, have each grade level provide a short overview of that concept and what it looks like at that specific grade level.

4. Consider having a person record the big ideas on a piece of chart paper for reflection. An idea for recording the vertical connections is shown below.

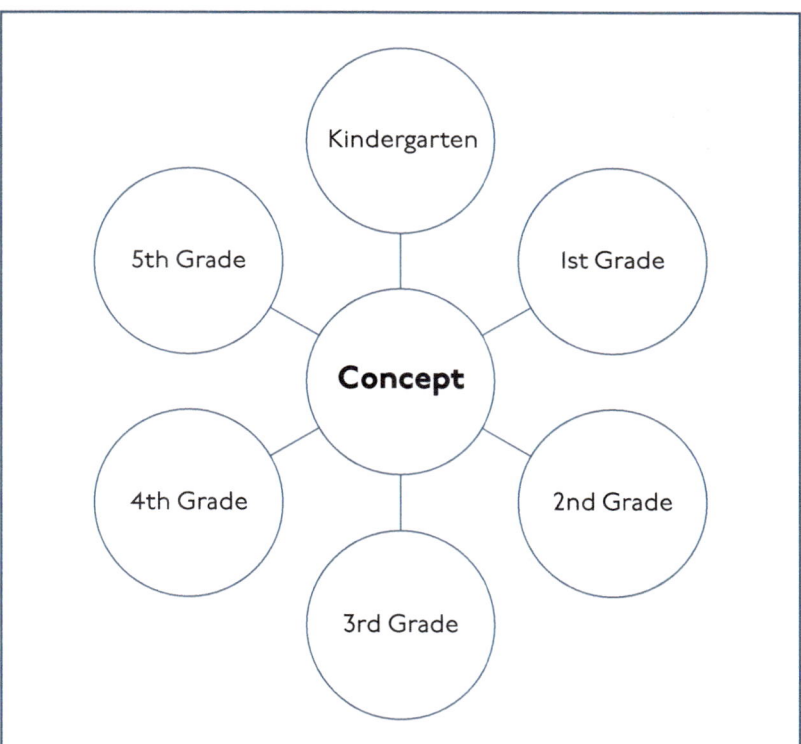

5. After sharing, discuss with your colleagues the following reflection questions:
   - What connections are you noticing between grade levels?
   - How do the concepts build and work together?
   - What new ideas or "aha" moments are you having?

Let's Try It (10 minutes)

Connections among ideas, concepts, and representations in mathematics are everywhere, but educators need to consider how they make time for students to see and verbalize the connections. In addition, students need to have opportunities to think about how mathematics is connected to them personally. Where do they see mathematics? How do they use it outside the four walls of the classroom?

With your teaching community, select one of the data collection options below. Bring your classroom data to next month's meeting to discuss with your colleagues.

Record this task and deadline in your calendar or planner.

OPTION 1: WHAT CONNECTIONS DO STUDENTS MAKE?

Determine when you will introduce a new concept to your students. Before providing any instruction, ask your students to tell you everything they know about the new concept. For example, when getting ready for a new unit on geometry, a fifth-grade teacher might show students a visual of a three-dimensional box and ask students to tell them everything they know. Record their ideas. Take a picture of the class chart to discuss with a colleague the connections students made.

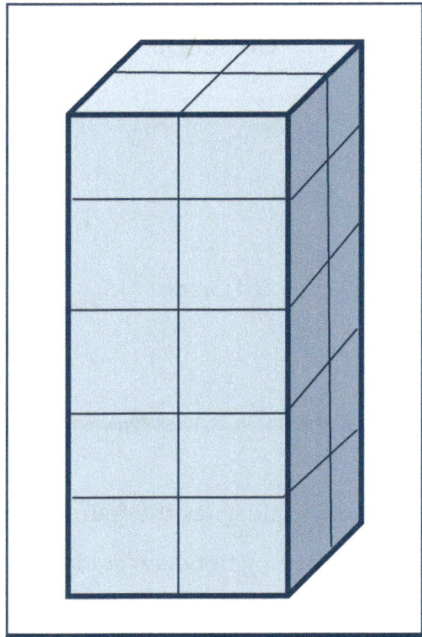

OPTION 2: LOOKING FOR VERTICAL CONNECTIONS

In your planning for the upcoming weeks, focus on one specific lesson. Write down the content that students will be learning. Look back to the content from the prior grade level to determine what connections you can make to students' prior learning. You may even check in with a teacher at the grade level below you. Use the following questions to guide your exploration:

- How does the content change from the grade level below you to your current grade level?

- What strategies are introduced to students that may be helpful to connect to?

- What vocabulary have students been introduced to that might be helpful in your instruction?

- What question(s) could you ask students to help them connect their learning from last year?

OPTION 3: CLASSROOM DISCUSSION

Tell students to think of things they do at home every day, such as brushing their teeth, eating meals, or playing with toys. Record their ideas on a class chart. Next, have students brainstorm how their everyday activities involve mathematics, for example, counting the time for how long they brush their teeth or naming the time they eat different meals. Take time to reflect on student responses, and note how they are connecting mathematics to the world around them. Tally the number of responses that students were able to connect mathematically. Share the findings with a colleague at next month's meeting.

ADDITIONAL PROFESSIONAL LEARNING EXPERIENCES

This month's additional professional learning experiences engage educators in experiences related to volume. Moving from one- and two-dimensional measures, volume can be challenging not only to visualize but also to conceptualize (Confer, 2017). Spending time making connections between area and volume can be especially helpful in making sense of three-dimensional figures.

Math Talks: Which One Doesn't Belong?

This month's math talk is a great way for students (and educators) to see connections between concrete, representational, and abstract mathematical representations.

1. Use four connected problems. There should be a relationship between each of them while showing some differences.

2. Ask the group to consider which of the four does not belong and to think about their justification for why. Challenge the group to find a reason for more than one of the problems.

With your math teaching community, give the volume problem below a try. Ask, "Which one doesn't belong? Why do you think so?"

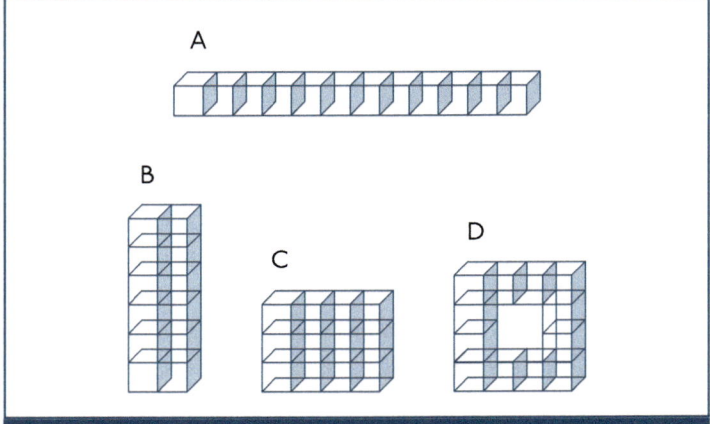

online resources   Available as a downloadable resource on the companion website.

Discuss the experience:

*   How did the Which One Doesn't Belong routine help you to see connections among the different representations?

*   How did this help you to think about and experience volume differently than you may have before?

*   How might you use the Which One Doesn't Belong routine in your classroom?

More ideas for Which One Doesn't Belong can be found on the website (Bourassa, 2013; https://wodb.ca/).

Manipulatives and Models Matter: Everyday Objects

As we continue to build our understanding of the connections that math has to ourselves and the world, we can invite our world into the classroom. Using everyday objects can help us to see the mathematics that is everywhere. Consider collecting a variety of objects that have volume such as dry beans, Starbursts or jellybeans, sugar cubes, toy cars, or dice. Next, choose a container that is large enough to hold multiple of your objects and include a scoop such as a small drinking cup or measuring cup.

For example, a clear food container is great for dry beans.

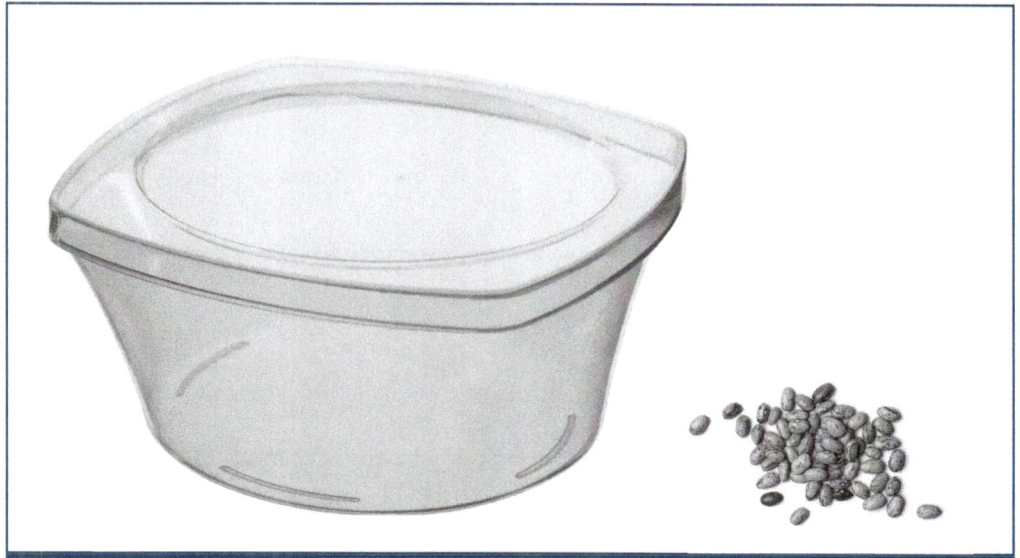

Image sources: Istock.com/Tohid Hashekmhani; Istock.com/PicturePartners

The big ideas about volume can be uncovered when we have multiple meaningful experiences. The act of estimating, filling up, counting, and comparing helps learners make sense of what it means to have volume and how we can find volume.

Try this experience with your math teaching community.

Working in pairs or small groups, select a container, scoop, and object to fill your container. Start by estimating how much of your item it will take to fill your container. You might use the scoop to take a sample to count and support your estimate. Record your estimation. Next, devise a plan with your group on how you will figure out the total

(Continued)

(Continued)

number it takes to completely fill your container. Once you've determined your plan, carry it out to find the total number of objects that will fill your container.

As a whole group, debrief using the following discussion questions:

- What strategies did you use to determine the volume of your container?
- How did this experience uncover big ideas around volume?

REFLECT

- What connections does this task have to your mathematical experiences outside the classroom?
- How might you modify this task for your own students?
- How might students approach this experience?

Game Time

This month's game offers students the opportunity to connect math tools to math concepts by having them create a representation of volume. You may choose to have students play competitively and see who can meet the challenge of building five different-sized skyscrapers in five rolls, or you can play cooperatively and engage the learners in collectively meeting the goal.

BUILDING SKYLINES: COOPERATIVE GAME

Materials:

Any number of players

Connecting Cubes

Dice: 3 per team

Paper

Pencil

Challenge: Create a skyline of buildings that are different heights.

Game directions:

- Roll the 3 dice and record the numbers rolled on your paper.

- Determine which two numbers will create your base of the building and which will represent the height.

    ○ For example, 2, 4, and 6 are rolled.

    ○ 2 and 4 can be used to build the base.

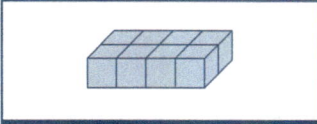

    ○ 6 will be used to build the height.

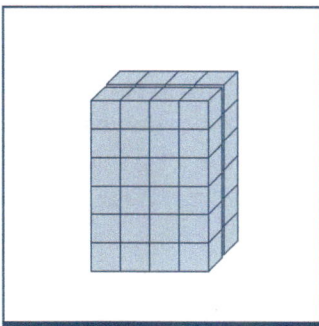

Build your skyscraper to match the dimensions you rolled.

- Record the equation that represents how to find the total volume.

    ○ For example, 2 × 4 × 6 = 48

- Repeat Steps 1–4 until you have 5 skyscrapers of different heights.

Discuss the experience:

- How does this cooperative game help students to solidify their understanding of volume?

- How are students connecting different representations of volume?

- What math habits would students use to play *Building Skylines*?

- What other ideas do you have for competitive gameplay?

online resources ☞ Available as a downloadable resource on the companion website.

IN YOUR CLASSROOM

Classroom Story

Mrs. Whitetail is a first-grade teacher who wanted to engage students in a task that would allow them to apply all of their new learning from the unit. She was particularly keen on noticing the connections students were making between tools, concepts, and their own lives.

She began her lesson by reading Thunder Boy Jr., *a wonderful story written by Alexie (2016) about an Indigenous boy who receives his given spirit name, a practice carried out by some Indigenous people. After the story, she posed a couple of questions to her class:*

- *Thunder Boy Junior was given a new name much longer than his birth name. How many letters long is it?*

- *How many letters do you suppose are in our whole class's first names?*

Mrs. Whitetail's students started by making estimates and then took connecting cubes to represent the number of letters in each of their own first names. Students used their towers of connecting cubes to compare their quantities.

Mrs. Whitetail then asked students, "What might be a way that we could find the total number of letters?"

Tommy:	*We could count them all.*
Mrs. Whitetail:	*I see many of you agree with Tommy. What would that look like?*
Tommy:	*We could start with my stack and then go around and just keep counting up.*
Mrs. Whitetail:	*Is there another way we could find the total?*
Masha:	*Well, I have 5 cubes and so does Mason. If we put them together that would make 10.*
Mrs. Whitetail:	*How might that help us count the letters in our names?*
Mason:	*We could see who else has two fives and put them together.*
Masha:	*And then we could count the groups of ten. We've been counting by tens!*
	Mrs. Whitetail was thrilled to see that her students were connecting their work of making tens and counting by tens to the name task. Several students showed their stack of cubes and were able to connect their stack with others to make groups of ten. Once there were only small stacks left, Ravi raised his hand.
Ravi:	*Is it ok to put three stacks together?*

Mrs. Whitetail: Tell us what you mean.

Ravi: Well, I have 4 and Sarah has 5 and Ian only has I left in his stack.

Mrs. Whitetail: What does everyone think?

Students were excitedly shaking their heads. The class finished putting stacks together and Mrs. Whitetail lined the stacks of ten up for students to begin counting. She then wanted students to see their class data in another way. Each student was given a sticky note and asked to write the numeral that represents the number of letters in their first name. Mrs. Whitetail drew a line on the board and wrote the numbers from 3 to 12 underneath. She modeled for students how to bring their sticky notes up to create a line plot of the frequency of letters in their names. Once students arranged their sticky notes, the class was asked to think about what they noticed.

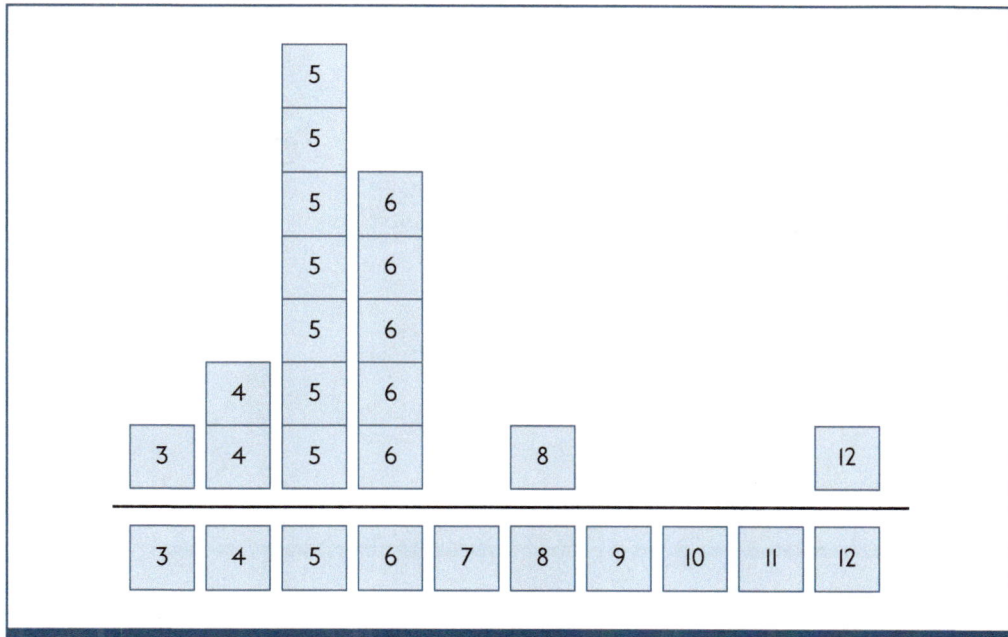

Jacob: This looks like the thing we used last week (pointing to the number lines they had been working on).

Mrs. Whitetail: It sure does. Tell me how you see the number line.

Jacob: There are numbers on the bottom that go just like the number line. Except there isn't 0, 1, or 2.

Mrs. Whitetail was pleased that many other students were agreeing with Jacob. Students continued making great connections from the cubes to the line plot and discussing the groups of ten that could be seen in the numbers. Students even began wondering what a line plot would look like if they had used their last names. Mrs. Whitetail knew that because students were making connections, this new learning would stick and students would be able to apply these understandings in new situations.

Anchor Lesson

Mathematics Is Connected

WHAT?

This lesson provides students an opportunity to make connections between mathematical concepts. Students will create a mind map with their math community based on their current background knowledge. An example is shown below.

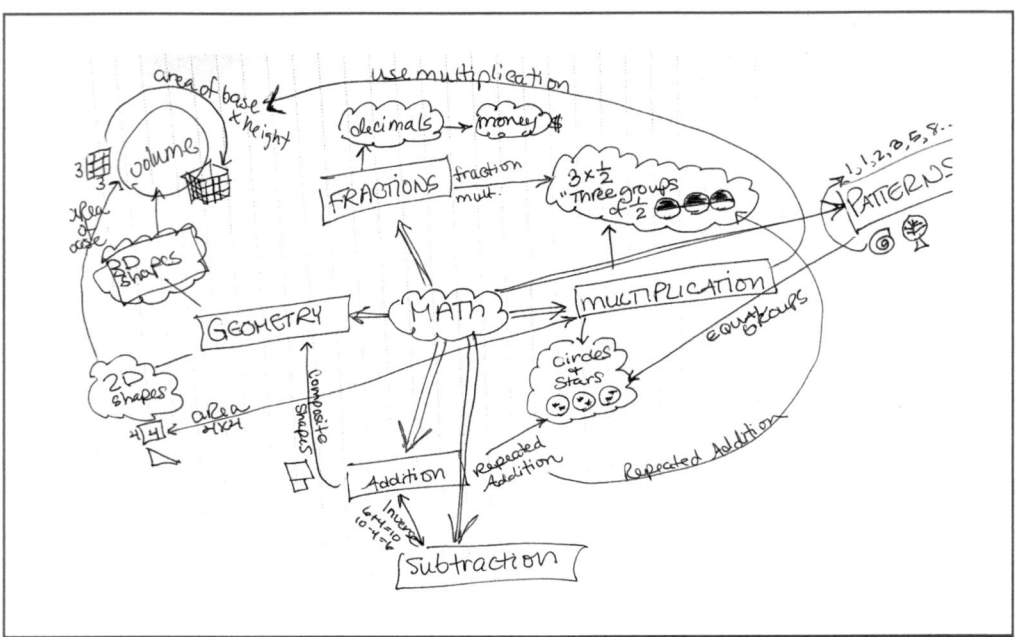

YOU NEED:

Chart paper (2)

Markers

HOW?

1. Write "Making connections" at the top of an anchor chart.

2. Display a second piece of chart paper. Engage students in creating a mind map of connections between mathematical ideas. Discuss with students the notion that the mathematical ideas they learn are all connected. No one idea sits alone. Tell students they will work on creating a representation of the connections within mathematics.

3. Write three words representing mathematical ideas or topics on a piece of chart paper. For instance, with a class of fourth- or fifth-grade students, you might write the words "addition," "multiplication," and "geometry."

4. Ask students to talk about how these ideas are connected. Show the connections by

drawing arrows from one idea to another, and record students' ideas about how they are connected.

5. Ask students to suggest additional mathematical ideas to add to the mind map. Record these ideas along with the connections that students describe. Continue making connections until students have exhausted their ideas.

6. Consider hanging the chart in a place where students can see and reference it. When new ideas are introduced in math lessons, students can think about how to add these new ideas to the chart and how to connect them.

7. Start a classroom discussion about the importance of making connections in math. Record their ideas on the anchor chart "Making connections".

REFLECT

Our students are better able to apply their learning when they have a deep conceptual understanding of math. We can support them in building a strong understanding of concepts by providing time for students to make connections between mathematical ideas, different strands of math, and from previously learned concepts to new concepts.

Use the following discussion questions to reflect with students:

- What new connections did you make?

- How do connections between different mathematical concepts support your learning?

- How might the mind map change if we added in connections from math to the real world?

LET'S DO SOME MATH ACROSS OUR SCHOOL!

In building a powerful math community, individuals need time to consider how their lives are mathematical. When we see connections from ourselves to the world around us, we see the mathematical power that we have.

Tell students they will get an opportunity to build their personal concept maps to show their personal connections to mathematics. Create and share your example to spark students' ideas.

1. As a class, take some time to brainstorm ideas about how mathematics is used in different areas of life. Consider creating headings such as hobbies, toys, sports, chores, or daily routines. As students share ideas, they should describe how these ideas involve mathematics. A student who offers the idea of playing the piano, for instance, might talk about the fractional values of notes on a piece of sheet music or the patterns of black and white piano keys.

2. Provide each student with a large sheet of paper and have them write their name in the middle.

3. Allow students time to create their concept map by recording and describing their personal connections to mathematics. Younger students can draw pictures of things they do that are mathematical. Older students might create a PowerPoint presentation using images that depict how they use mathematics in their own lives.

Post your students' concept maps outside the classroom where others can admire and celebrate your students' math learning.

TEACHING MOVE: PLANNING CONNECTING QUESTIONS

Helping students to see the connections between math concepts and representations requires teachers to intentionally plan the questions they will use during instruction. During your math planning, take some time to scan the list of questions and write one to two questions in your teaching plans. Here are some examples of questions to support students in making connections.

QUESTIONS TO CONNECT CONCEPTS:

Where did we use [math concept] in our work today?

What mathematical ideas have we learned about in the past that we used today in our work?

How is [prior concept] connected to [new concept]?

QUESTIONS TO CONNECT REPRESENTATIONS:

How does this equation represent the model you made?

Where does your model show [math concept]?

How does your answer connect back to the original problem? Does it make sense?

If you were to describe to someone what the numbers and other representations in your problem stand for, what would you say?

How does [student's] representation connect to [student's] representation?

REFLECT

- How might intentionally preplanning questions to connect concepts help students see the connectedness in mathematics?

- How might students' learning be supported when you help them connect representations?

TEACHING MOVE: TIME FOR REFLECTION

In the day-to-day experiences within the classroom, it can be challenging to make time for the in-depth processing and analysis required for students to know concepts deeply and to be able to apply them. As educators, we must be intentional about the time we set aside for the essential practice of reflection so that students can think about their learning from the day and make solid connections.

APPLY

- Commit to providing time at the end of each lesson for reflection on learning, even if students have not yet finished their work. As you begin a lesson, set a timer to go off about 10 minutes before your math block is set to end.

- When the timer goes off, gather your students together for reflection. The following discussion questions could be used to reflect on the learning and help students to make connections:

    ○ What did we learn today? What prior knowledge was helpful in your learning today?

    ○ What representations did you use today? How have we used them before?

    ○ Where have you seen what we did today outside the classroom?

REFLECT

- How might setting aside time for reflection help students make connections?

- What math habits might students be building when they reflect on their learning?

CLASSROOM ROUTINE: WHICH ONE DOESN'T BELONG?

The Which One Doesn't Belong math sense routine forces students to look at connections between equations, numbers, visuals, and other representations of mathematical concepts. Teachers can use this routine as a daily 5-minute warm-up to build student reasoning and communication. The goal is not to identify a correct answer but to grow students' ability to see connections among similar mathematical ideas and to be able to justify their thinking.

Implementing the routine:

1. Select a group of four related problems or images.

2. Ask students to think about which of the four doesn't belong with the others and why it doesn't belong.

3. Call on several students to share.

4. Encourage students to make arguments in support of more than one problem or image in the group.

Ideas for K–2:

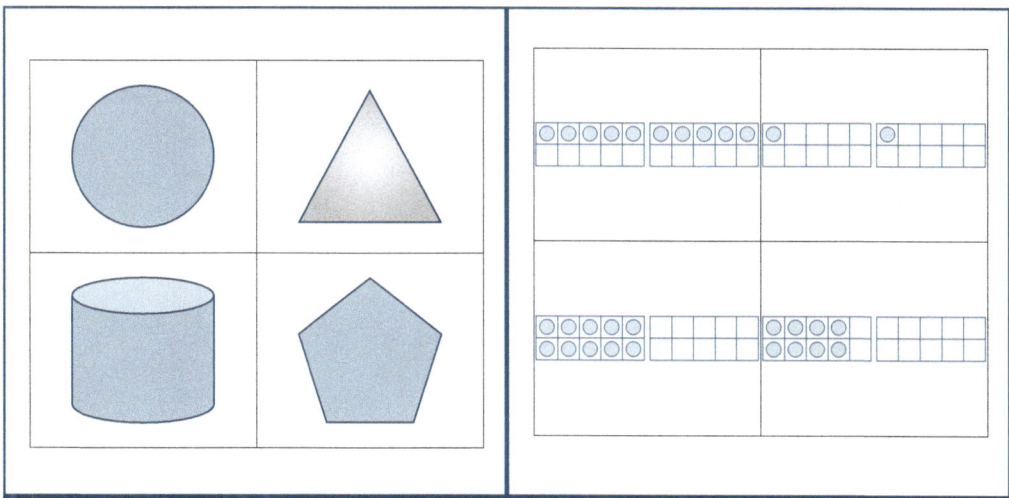

(Continued)

(Continued)

Ideas for 3–5:

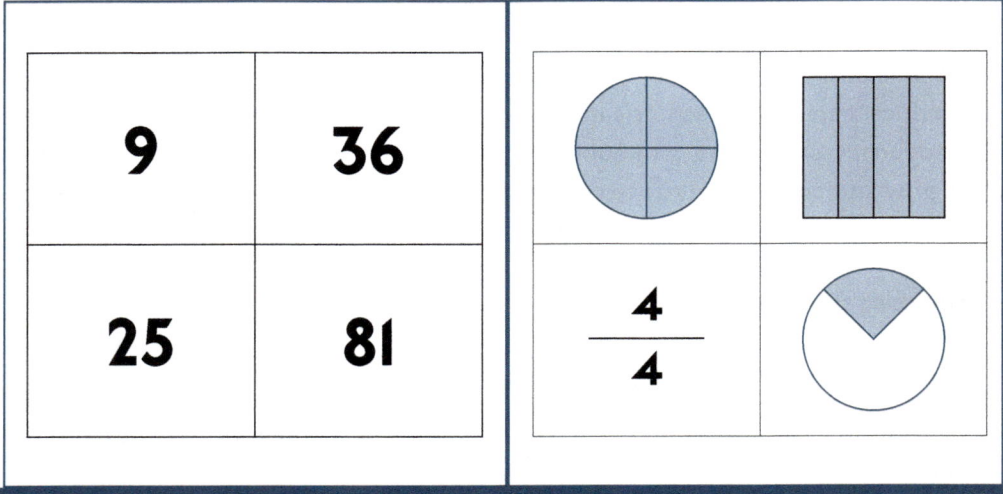

You can create your own images for Which One Doesn't Belong by using the math content you are currently teaching. Students can even create Which One Doesn't Belong sets of their own to share with their class.

Additional Which One Doesn't Belong resources can be found on the website **wodb.ca**

STATION ACTIVITY: CONNECTING MANIPULATIVES TO CONCEPTS

This month's station activity extends the work students did with manipulatives and the cooperative math game. Students will explore three-dimensional shapes and the properties that connect area to volume. Although young students are not yet learning about area and volume formally, they are using the math tools to think about the general concept.

GRADES K-1

In kindergarten and first grade, curriculum standards have students explore three-dimensional shapes. They are starting to define attributes of shapes and to think about combining shapes to make other shapes. In this activity, students will use a target number of cubes to build as many different buildings as possible.

Materials: connecting cubes (12 per student)

Pose the problem: How many different ways could you make a building from 12 cubes?

You might start with a simpler problem. Show students 4 cubes and have them consider what a building might look like that is made from 4 cubes. Students could lay the cubes in a single row of 4, or create a 2 × 2 × 1 prism.

GRADES 2-3

Materials: connecting cubes (24 or 36 per student)

Pose the problem: How many different ways can you make a building from 24 or 36 cubes?

GRADES 4-5

Materials: connecting cubes (any number of cubes less than 50)

Pose the problem: A building company is working with the school district to determine plans for a new school building. There will be no more than 48 rooms in the building. Choose a number of rooms and a configuration to present to the company and school

(Continued)

(Continued)

district. With each connecting cube representing one classroom, record the volume of each floor in your building and the total volume of the school in cubic units.

For example, with 36 connecting cubes, students could build a school that has 12 classrooms on each of 3 floors. It would look like the following.

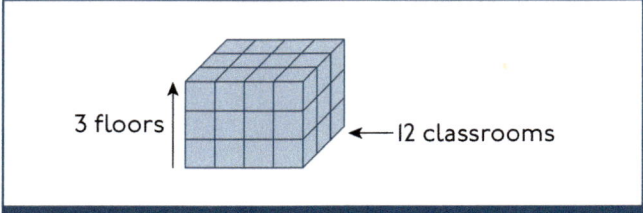

Students could justify their configurations by writing about their reasons for constructing the buildings in the ways that they did.

 Available as a downloadable resource on the companion website.

LITERATURE CONNECTIONS

***Look, Grandma! Ni, Elisi!* By Coulson** (2021) (Grades K–2)

Bo needs a container to hold his beautifully decorated marbles to sell at the Cherokee National Holiday festival. A pot is too small. A wooden tray is too big. A vase is too tall. Bo uses his understanding of volume and makes lots of mathematical connections as he solves his real-world problem.

Activity: Find a container to hold a collection of objects that is important to you. What math connections did you make as you figured out how to fit your objects into a just-right container?

***Missing Math: A Number Mystery* by Leedy** (2008) (Grades 3–5)

One day all numbers mysteriously disappear, resulting in a multitude of real-world math problems that can't be solved. Fortunately, the cause is identified, a number vacuum designed to create infinity. The vacuum is put in reverse, the numbers are restored, and everyone has a new appreciation for the value of numbers.

Activity: What would happen if numbers disappeared from your life? What are some activities you would no longer be able to do? Write a thank-you letter to math for the ways that it helps you to do things that are important to you.

SPOTLIGHT ON BRAIN SCIENCE

Memory plays an important role in learning. We have three types of memory: long-term memory, working memory, and short-term memory. When we know how each type of our memory works, we can more effectively remember what we learn and retrieve needed information from memory.

You can share the following information with students in a mini lesson or a station activity. You might also send this reproducible home along with the Family Newsletter to help families talk about these important ideas.

CONNECTIONS HELP US REMEMBER WHAT WE LEARN.

Brain scientists have discovered three types of memory.

1. Short-term memory: When you learn something new, your brain holds this information in short-term memory for 10 seconds or less. In these 10 seconds, your brain decides if it will move this information to your working memory or get rid of it.

2. Working memory: Your working memory is at work when you are thinking. As you think about your new learning, you are helping your new learning to move from your working memory to your long-term memory.

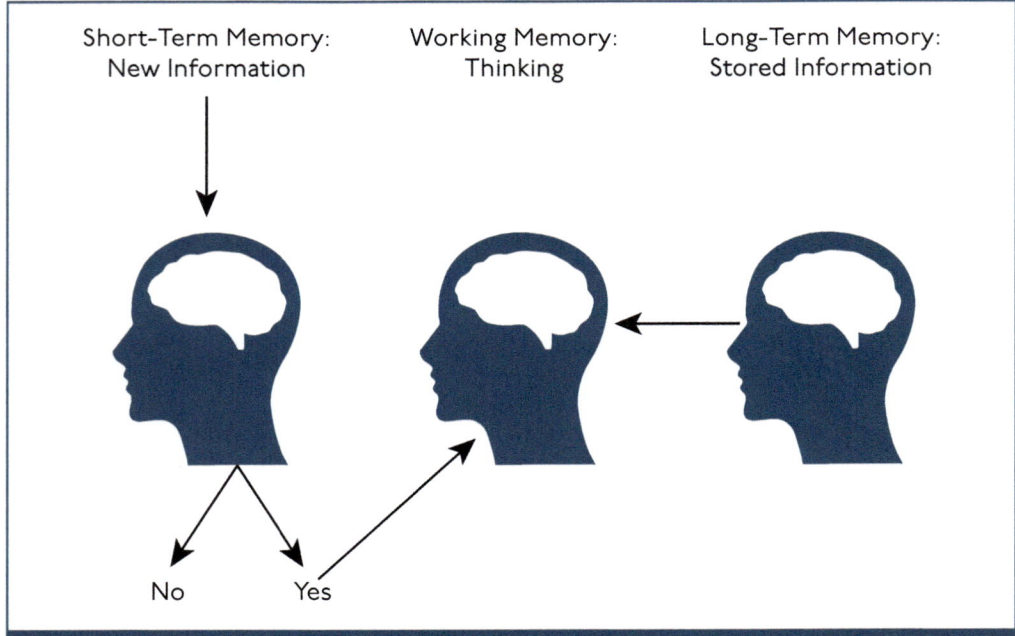

Image source: istock.com/Vedc76

3. Long-term memory: Your brain stores information that you remember in long-term memory. You want to store the things you learn in long-term memory so you can use this information when you need it. When you need to use something you have learned, your working memory finds this information in your long-term memory and puts it to work.

Take a look at the images. Think about what you notice and share with a partner.

HOW YOU CAN HELP YOUR MEMORY

1. When you learn something new, do something with your learning right away. You can:

   - Tell someone what you just learned.

   - Write down what you just learned.

This forces the new learning into your working memory.

2. When you learn something new, think about how it is connected to other things in your long-term memory.

   - Ask yourself, "What do I already know about this?"

   - Make connections from your new learning to things you know. Tell someone else about your connections.

Choose one of the following ways to think more about this important information:

1. Write down why these ideas are important to you and how you want to put them to work in your life.

2. Talk to a friend about these ideas. Decide together how you will put them to work.

3. Share these ideas with someone at home. Tell them why these ideas are important to you and how you plan to use them.

 Available as a downloadable resource on the companion website.

SPOTLIGHT ON EQUITY: HUMAN CONNECTIONS STRENGTHEN MATH LEARNING

Boaler stated that our students' abilities to make mathematical connections are related to their abilities to collaborate with peers:

> Mathematics is often depicted as the most solitary of subjects, but it is a discipline, like all others, that has been built through connections between ideas. New ideas and directions come from people reasoning with each other, setting out ideas, and considering the ways they are connected to each other. (2019, p. 200)

In other words, human connections support connection-making in math, and both of these abilities enhance learning. The ability to connect and collaborate with peers is needed in a math classroom that is both challenging and caring. It is prerequisite to success in school and in life. The development of collaborative abilities is, therefore, essential to equitable outcomes in school.

Boaler explained that effective collaboration creates "relational equity" (p. 186):

> One of the goals of schools should be to produce citizens who treat each other with respect, who value the contributions of others with whom they interact, irrespective of race, class, or gender, and who act with a sense of justice, considering the needs of others in society. A first step toward producing citizens who act in such ways must be the creation of classrooms in which students learn to act in such ways, for we know that students learn a lot more than subject knowledge in their school classrooms. (pp. 186–187)

Boaler suggested three mindframes teachers need to cultivate in students to build their collaborative abilities:

1. Open Minds—When students value human differences, different perspectives, and different strategies, they are open to learning from others' ideas.

2. Open Content—When students understand that mathematical ideas can and should be represented in multiple ways and that there are multiple correct strategies for solving problems, they are open to considering new ways of thinking.

3. Embrace Uncertainty—When students understand that productive struggle leads to learning and see mistakes as valuable, they recognize uncertainty as a natural part of the learning process and see social interaction as a support for moving through uncertainty.

TRY IT

Observe your students as they work on math in collaborative settings. Jot down things that students say and do to:

- Demonstrate that they value each other

- Demonstrate that they value different ways of thinking

- Help each other make connections between mathematical ideas

Share these exemplars of effective collaborative behavior with your class. Ask students to talk about why these behaviors are important. Together, set a small goal and create a plan for strengthening the class's collaborative abilities.

MATHEMATICAL ME: STUDENT JOURNAL AND PORTFOLIO

Student Journals:

Have students respond to the following questions:

- How does making math connections help me learn?

- How can math connections help me solve new problems?

Older students can write their responses in their Mathematical Me journals. Younger students can draw a picture showing their response, or you might gather responses during a class discussion or quick one-on-one interviews.

(Continued)

(Continued)

Review students' responses to monitor students' beliefs about mathematical connections. You might summarize this data by tallying or graphing the different ideas students offer.

Be sure to share and discuss this data with your class. Ask, "What does this data tell us?" and "What are some things we can do to continue growing as mathematicians?"

Student Portfolios:

Have students choose a piece of work where they made a math connection. On a sticky note, have them write the date and complete this sentence frame:

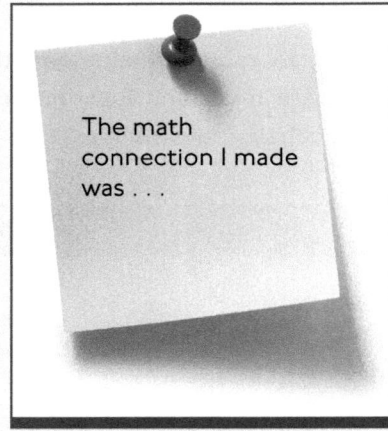

The math connection I made was . . .

Source: Istock.com/Thammask-Chuenchom

FAMILY NEWSLETTER: WE MAKE MATH CONNECTIONS!

WHAT WE'RE LEARNING

This month we are learning about the different ways we make connections in math and how these connections support our math learning.

WHY IT'S IMPORTANT

When your child makes connections between math concepts and representations, they deepen their understanding of the mathematics they are learning. When your child intentionally connects mathematical ideas, they remember what they learn because their brains build strong pathways that allow them to access their learning when needed. When your child sees how math is connected to real life, they recognize why their math learning is important.

HOW YOU CAN HELP

Here are some ideas to help your child make mathematical connections:

I. Math is everywhere. Talk with your child about how you use math at home or work. You might discuss how you use math in activities such as these:

- Paying bills
- Managing time
- Organizing groceries on shelves
- Measuring

2. Encourage your child to think about math beyond just numbers. When you help your child with homework, have them tell you about what they are thinking. Ask them, "Could you draw a picture to show what you are thinking?" When your child represents their math thinking with pictures and words, they are making important connections that support their learning.

online resources ⟋ Available as a downloadable resource on the companion website.

CHECKING IN ON OUR LEARNING: WE MAKE CONNECTIONS!

- How does making connections help me learn?

- How can math connections help me solve new problems?

This month's learning focus, making connections, moves students away from the idea that mathematics is a bunch of unrelated facts and skills and is only done in school. Learning is supported when students see how each newly discovered mathematical idea is connected to something they already know. The connections they make strengthen their understanding of the mathematics taught in school, positioning students to begin thinking beyond concepts learned in the classroom. Armed with a strong understanding of concepts and the knowledge that mathematics is everywhere, our students become agents in their own math learning, uncovering mathematical beauty and solving real-life mathematical problems.

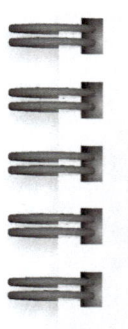

MATHEMATICAL ME: EDUCATOR JOURNAL

- How has this month's learning focus supported your students' mathematical growth?

- How has this month's learning focus supported your growth as a math teacher?

- How has this month's learning focus supported your school's growth as a powerful math community?

LOOKING AHEAD

In April, we will look at a variety of types of mathematical patterns and consider ways that mathematicians use patterns to make sense of the world, generalize mathematical ideas, and solve important problems.

Questions for Reflection

- How do patterns show up in your life?

- What types of mathematical patterns are you familiar with?

CHAPTER TEN

April
We See and Use Patterns!

ESSENTIAL QUESTIONS

- How does looking for math patterns help me learn?
- How can I use math patterns to solve new problems?

THIS MONTH'S FOCUS

In a powerful math community, leaders, teachers, and students notice and use patterns to make sense of not only mathematical ideas but also of the world around them. Math learners look for patterns in numbers, shapes, or ideas and use those patterns to generalize mathematical ideas. They are also aware of patterns in their life and work, and they see the beauty of patterns in the natural world.

Educators help students to experience mathematics as the study of patterns. They nurture students' natural inclination to seek and use patterns because they know that this habit lies at the heart of what it means to be a mathematician.

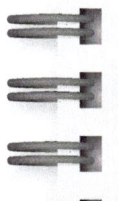

MATHEMATICAL ME: EDUCATOR JOURNAL

Take five minutes to respond to the essential questions. You will have a chance to reflect on these same questions and to have students respond to these questions at the end of the month.

ON YOUR OWN

LET'S DO SOME MATH!
VISUAL PATTERNS

The X™ (formerly known as Twitter) math community is a great place to connect with other mathematics teachers, leaders, and learners. Howie Hua is one of my (Holly) favorites, and he is a great example of finding the fun and beauty of mathematics in everyday life and the world. In addition to creating easy-to-follow videos on mathematical algorithms and why they work, he uploads everyday photos that depict mathematical patterns. By asking his followers, "How many did you see and how did you see them?" the X community gets exposure to how others think mathematically.

Take a look at one of his posts below. Consider how many tiles you see and how you see them. How many different ways can you figure out the number of tiles?

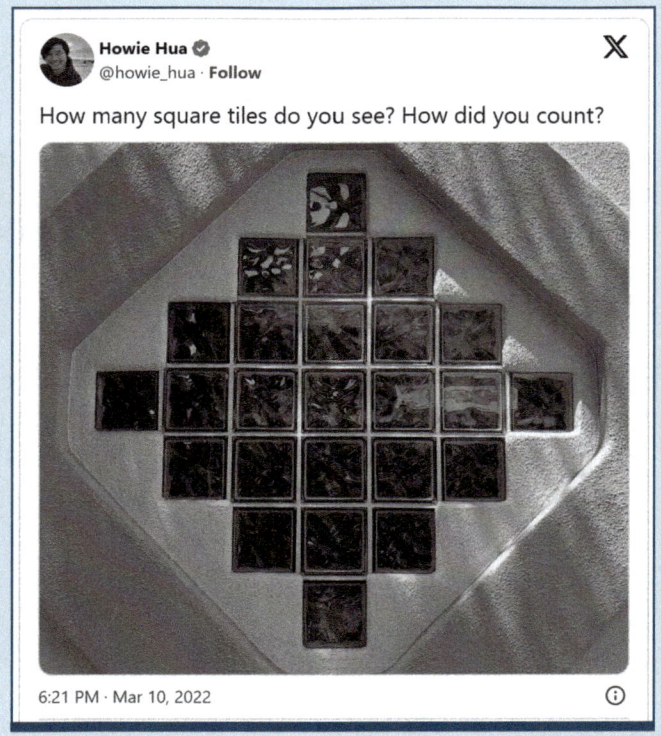

Source: Reprinted with permission from Hua (2023, January 15).

Record the different ways you solved this problem to share with your colleagues. How did you engage with the habits of mathematically powerful people as you thought about and solved this problem? You may even take a photo of another visual pattern that you see to share with your colleagues and your students.

_____

 Available as a downloadable resource on the companion website.

Why This Focus

Holly's Math Story:

I've realized that I am a natural pattern noticer. I am constantly curious about things that happen again and again, often sharing my wonderings with my husband. He frequently calls me _Curious George_ and chuckles at my childlike wonder of things around me. But for me, I think there's something wonderful and comforting about unlocking a mystery of the world around me simply by noticing things that happen again and again. I love the moment I realize it's spring when I sit on my front porch with coffee and I start to hear the chirpings of baby birds. The noises tell me that winter is about to thaw and warmer weather is on its way.

I have come to believe that my curiosity, wondering, and noticing about the patterns around me are what makes me so interested in the field of mathematics. Mathematicians frequently define mathematics as a way of understanding the world through the study of patterns (Boaler, 2015; Gonzalez, 2023; Woo, 2019). If the heart of mathematics is to seek out, identify, and use patterns, we must consider how we engage students in this habit. Educators need to expose students to the variety of ways in which patterns emerge in math.

We can think about several ways that mathematical patterns show up in the world around us. Patterns exist in nature and the landscape of our world. Patterns describe how the world functions in our human-made spaces. Being human and engaging in human behavior is a series of patterns. In addition, patterns play a pivotal role in helping learners think about mathematical concepts and connect known to new concepts.

Patterns Exist in Nature

I have become fascinated by the various shapes of succulents and how their intricate patterns are mesmerizing. They have a repeated element

> If the heart of mathematics is to seek out, identify, and use patterns, we must consider how we engage students in this habit. Educators need to expose students to the variety of ways in which patterns emerge in math.

and a sort of balance in the way their leaves spiral out from the center. Where else do we see patterns in nature?

Images source: istock.com/kynny

Boaler (2015) wrote extensively about the beauty and wonder of mathematics. She described how mathematics shows up in nature when we see the spiral of a seashell, which has a visual pattern and can be represented numerically with a special pattern called "Fibonacci numbers." These types of patterns show up frequently in nature. The hexagonal patterns of the snowflake are consistent, and even animals, such as the dolphin, use a pattern of clicking noises to alert their friends of their location through the mathematics of rate.

Terence Tao is a mathematician and the Fields Medal winner from 2006 who has spent years uncovering patterns in prime numbers. When asked about mathematics in the universe, he stated:

> At the extremely microscopic level, the laws of nature are ordered. Particles and quantum waves obey very rigid waves of mechanics. But as you go to more complicated objects, molecules and living creatures, then it becomes more chaotic and unpredictable. There's this weird mathematical phenomenon called universality. You get very complicated systems, of atoms or people, but if you look at it at a large-enough scale, order starts emerging. Einstein once said that the most incomprehensible

thing about the universe is that it is comprehensible. It is very complicated, but at certain levels, patterns appear again. (quoted in Bernstein, 2019, para. 47)

When we return to the definition of mathematics as a way of describing the world around us, the natural world exposes its patterns to us only if we take the time to notice.

Patterns Exist in Our Human-Made World

People use patterns in work and daily life. Drivers use the knowledge they have internalized about patterns to successfully navigate to their destinations (Woo, 2019, p. 7). Scientists use patterns in data to seek out ways to stop the spread of cancer. Financial advisers pay attention to market patterns that yield positive results for their clients (Gonzalez, 2023).

We all use patterns to understand, communicate about, and use things we notice. When we teach students how to build their skills in seeking out and analyzing patterns, we provide them with important life skills for their futures. As educators, we open up the world for students when we teach them the math habit of pattern-seeking.

Patterns Exist in Human Behavior

Humans are inherently pattern seekers. Woo said that we are born to look for and make sense of patterns:

> The human brain is nothing if not a pattern-recognizing machine, built from the ground up to perceive patterns in our surroundings. You can describe virtually every human function of the brain in terms of its relation to patterns. What is smell? It's our recognition of specific olfactory patterns and associating some of them with good (sweet) and some of them with bad (bitter). What is memory? It's the connection of patterns with specific meanings, the facial and vocal cues of people we meet whom we can therefore later recognize. (2019, p. 6)

Imagine you are in unfamiliar surroundings. First, you seek out what is known, recognizing patterns of familiarity. For example, you pull up to an unfamiliar grocery store to do your shopping. Your brain may quickly seek out where the carts are located to start your shopping experience. As you enter the store, you examine the layout, determining where the checkout is located, and using your prior knowledge to identify sections of the store. Your past experiences may have taught you that fresh food is more likely to be found around the perimeter of the store while canned goods and boxed foods tend to be located in the middle of the store. The patterns you have recognized lead you to make informed decisions.

Using Patterns to Learn Mathematics

Math learners know that seeking out and using patterns is the foundation for uncovering and connecting mathematical concepts. In his publication, *Mathematics for Human Flourishing*, Su wrote:

> Learn a bunch of separate mathematical facts, and it is just a heap of stones. To build a house you have to know how the stones fit together. That's why memorizing times tables is boring: because they're a heap of stones. But looking for patterns in those tables and understanding why they happen - that's building a house. And house builders perform better in mathematics; data show that the lowest-achieving students in math are those who use memorization strategies, and the highest-achieving students are those who see math as a set of connected ideas. (2020, p. 38)

As educators, we can help students make sense of mathematics by intentionally setting aside time to find and analyze patterns, generalize from patterns, and communicate about these generalizations.

Seeing Patterns Among Mathematical Concepts

Teaching and learning math is enhanced when we see the connectedness among mathematical concepts. Taking time to look for patterns across different math domains supports mathematical understanding. Students who understand the concept of equivalence in number relationships, for instance, can apply the idea of equivalence in thinking about characteristics of geometric shapes. In her book *Faster Isn't Smarter*, Seeley (2015) wrote, "The patterns that underlie key mathematical concepts, such as equivalence or proportionality help students make sense of otherwise disconnected bits of knowledge" (p. 153).

- Consider two mathematical concepts you teach. What patterns do you notice between those concepts?

The Relationship of Patterns and Connections

Patterns and connections are closely intertwined. If you try to define one and then the other, you will find that their definitions tend to overlap. But what is happening is that there is a deep connection between the two. Patterns jump out at you. They tug on your sleeve and say, "Hi. I'm here. Notice me." When you stop to do so, your brain looks at the pattern, seeking to find a connection. It asks itself: How is what I see in front of me connected to something I already know? Our brains are wired to make these kinds of connections.

Patterns jump out at you. They tug on your sleeve and say, "Hi. I'm here. Notice me."

We may have needed this instinct in the Stone Age for protection in making quick decisions. Noticing patterns in animal tracks may have saved us from an untimely death. In our modern world, we notice patterns to seek comfort, order, and logic in the chaos. It may be an instinct that is ingrained in us all, but it is also the heart of learning and doing math. We look for patterns, seek to make connections, and look to find the logic in the chaos.

In this chapter, we will explore how patterns are at the heart of doing math. We aim to strengthen this habit and become pattern noticers as we build a powerful math community.

Tapping Into Your Experience

Patterns are not only found and used within the math classroom, but they also make their way into our lives, work, and hobbies.

Create a table like the one below. Think about how you have used patterns in mathematics, perhaps patterns involving shapes or numbers. List ways patterns are used to drive decisions in your day-to-day life, such as the route you take to drive to work. Think about how patterns are used in your work. For example, teachers recognize that the first day back from a long break is similar to the first days of school and, therefore, time to review rules and procedures. Lastly, think about your hobbies. Are you an artist? Musician? Gardener? How does your hobby lead you to use patterns?

Patterns I use in math:	Patterns in my life outside of work:	Patterns in my work:	Patterns in my hobbies:

After creating and reviewing your list, consider the following reflection questions:

- What role do patterns play in your life inside and outside of work?

- How does consciously thinking about the patterns you see and use help you in your role within your math community?

> In our modern world, we notice patterns to seek comfort, order, and logic in the chaos. It may be an instinct that is ingrained in us all, but it is also the heart of learning and doing math.

The Role of Patterns in the Math Habits

Mathematicians look for and use patterns. Students might be prompted to look across several connected problems and notice what is the same. Finding the connections between problems may help uncover a "rule" that can be used for other similar problems.

Consider the following series of problems:

4×10
4×100
$4 \times 1,000$
$40 \times 1,000$

Students could be asked what they notice about patterns in the factors and products. They would uncover the structure of multiplying by a power of ten by noticing patterns in the factors and products.

- What is a string of related computational problems appropriate for your grade level that could be used to uncover a mathematical generalization?

- How might this practice of finding and generalizing from patterns be useful for students learning math?

TOGETHER WITH YOUR TEACHING COMMUNITY

Since We Met Last (10 minutes)

In March, you and your colleagues gathered classroom data about connections that students make between mathematical ideas or from the math they are learning in school to mathematics outside the classroom.

You might use the following protocol to share and reflect on this data with your math teaching community:

- Share your data with a colleague (5 minutes)

- As a group, discuss what you learned from this data and how these insights can help you support your students' growth as mathematicians. (5 minutes)

Let's Do Some Math Together! (10 minutes)

In small groups, share the different ways you saw the tiles in Howie Hua's post (see p. 320). Show how you grouped the tiles, quantified the groups, and figured out the total number of tiles.

If you took photos of other patterns around you, share them with your colleagues.

Discuss:

- What patterns did you notice?

- What were some similarities between your strategy and your colleagues' strategies? Differences?

- What math habits did you use in finding the different ways and sharing with your colleagues?

- What are some other patterns you notice around you that would make a good problem to use with your students?

Building Our Expertise (30 minutes)

If we wish to become more aware of the patterns that exist between ideas, concepts, and ourselves and the world, we need to make time to notice and think about patterns. The patterns that we find help us to connect learning across content and problems for the benefit of our students.

In grade-level teams, take 15–20 minutes to look through two upcoming math units. Notice the problems that students will be solving, and write down any patterns you see across problems. This process will help you to unlock key concepts and consider the connections you will make from topic to topic. Use the following ideas as you scan through the content and resources:

(Continued)

(Continued)

- What big mathematical ideas do both units address?

- What types of problems will students encounter? Where might there be crossover or similar strategies in the problems from these two units? In other words, where might there be opportunities for students to notice patterns of ideas across these two units?

- How do the problems relate to the mathematical concept students are expected to learn in this lesson?

- What is a question you might ask students to help them notice the pattern of mathematical thinking within these related problems?

For example, Grade 4:

UNIT 2	UNIT 3
Multiplication - Area model - Place value and expanded notation	Division - Area model revisited - Inverse operations (how are multiplication and division related?) - Place value and expanded notation
Patterns: Both units use the area model starting with expanded notation using place value. Both units condense the area model and eventually stop writing expanded notation.	
Key takeaways: Emphasize place value and expanded notation for both units. Connect the relationship between the area model for multiplication and the area model for division.	

With your math community, engage in a vertical share about the patterns you notice across your curriculum. Starting with kindergarten, teams should talk briefly about a pattern they notice within their grade-level curriculum. Record the patterns that are shared. As a school community, consider the following questions:

- How can we strengthen student learning by helping students to notice patterns across math topics?

- How can we strengthen our math teaching by consciously thinking about patterns in our math curriculum across grade levels?

Close the session by asking your math community to reflect on this question: How can noticing patterns across the curriculum deepen students' understanding of mathematical ideas and their ability to reason mathematically?

Let's Try It (10 minutes)

This month, your team will focus on the practice of noticing patterns. Decide which route you will take from the options below and bring your classroom data to next month's meeting to discuss with colleagues. Record your data collection plan in your planner or calendar and bring the data to next month's meeting to share and discuss.

OPTION 1: CONNECTING PLANNING TO PRACTICE

With your math community, you worked to find patterns across math problems and math content in your resources. When you notice vertical patterns, you can connect students' prior learning to your grade-level content.

Select a single lesson from your resources that you will be teaching this month. Write down the big idea or focus for learning, important vocabulary, and representations students will use. Reflect on the vertical conversations you had with your team and take a few minutes to answer the following questions:

- What patterns do you see from your content and vocabulary to other grade levels?

- How can you use this pattern to help students make connections between content and prior learning to new learning?

OPTION 2: NOTICING PATTERNS IN THE WORLD

Show students an example or two of a mathematical pattern found in nature. A simple Google search yields some beautiful images. Challenge students to find an example while they're out on the playground, taking a walk with their family, or driving in a car. Set aside 5–10 minutes during math class for students to share what they found. Make some notes about the experience so you can debrief with a colleague at the end of the month.

OPTION 3: PATTERNS EVERYWHERE

On an anchor chart, create a table like the one that follows.

Patterns I use in math:	Patterns I use outside the classroom:	Patterns in my hobbies:

(Continued)

(Continued)

Discuss with students how they see patterns in math, outside the classroom, and in their hobbies. Create a list and save the anchor chart or take a photo of it to bring with you to next month's meeting. With your colleagues, discuss what partners students are noticing.

ADDITIONAL PROFESSIONAL LEARNING EXPERIENCES

This month's additional professional learning experiences involve pattern blocks, a set of colorful shapes used to think about patterns in mathematics. The versatility of pattern blocks makes them a powerful math tool to learn about and make sense of a variety of mathematical concepts. Pattern blocks consist of a yellow hexagon, red trapezoid, blue rhombus, green triangle, orange square, and tan elongated rhombus.

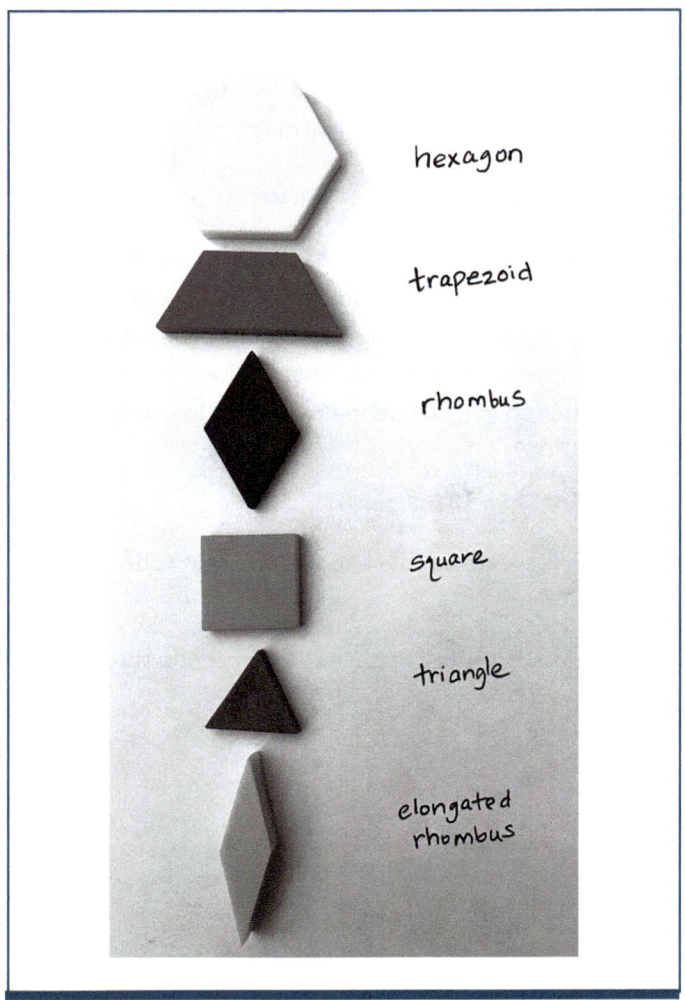

Math Talks: Shape Talks

This month's math talk is similar to the dot talk that was introduced in August and the number talks in November and February, but instead of numbers or dots, a design of shapes is used to talk about and quantify what is seen. Incorporating geometric shapes is a great way for all students to access patterns.

1.   Select a shape, series of shapes, or image built from different shapes.

   •   YouCubed.org is a great resource and has a section on their website with shape talks.

      ○   https://bit.ly/436OwzW

      ○   https://bit.ly/3TlSFg5

2.   Consider the mathematical direction you want the group to explore. For example, in K–I, you may choose to have students name a pattern that they see or name the properties of the various shapes. For older students, you may have students name quantities they notice within the shape, including fractional parts of the patterns.

3.   Ask a question that focuses on the mathematical concept you will explore and use a think–pair–share routine to generate ideas. Sample questions to ask are:

   •   What pattern do you see in the shape?

   •   What would you add/remove to make a new pattern?

   •   How can you use a pattern to tell how many ___ there are?

   •   How could you extend the pattern?

   •   What else do you notice?

4.   Share the ideas with the group.

With your math teaching community, try a shape talk with the image below. Ask, "What fractional name would you give to each shape below?" or "How did you use a pattern to help you?"

(Continued)

(Continued)

Discuss the experience:

- How did you use patterns to think about the fractional names for shapes?

- What fractional concepts did you use when thinking about this new shape?

- What math habits did you use while engaging in the shape talk?

- How might you use this routine in your classroom?

Manipulatives and Models Matter: Create a Shape

Pattern blocks are a great tool to represent patterning. Young students may use pattern blocks to create and extend patterns. In making an AB pattern, a hexagon and triangle might be used to demonstrate a simple pattern.

Students exploring two-dimensional shapes and their properties might sort the pattern blocks by number of sides and angles looking for patterns in their characteristics.

Squares, rhombi, and trapezoids would be placed in a group together all having four sides. In the example below, shapes are sorted by obtuse angles, acute angles, and both obtuse and acute angles.

As students get older, they may start to think about the relationship of one shape to another. In looking for patterns, they might notice that six green triangles cover one yellow hexagon; therefore, they may name the triangle one sixth when a single hexagon represents one whole.

Pattern blocks are a wonderful starting place to uncover big ideas regarding fractional relationships. Challenging students to use the pattern blocks to create designs that incorporate fractions allows them to apply their knowledge in a fun, creative way.

(Continued)

(Continued)

Try this experience with your math teaching community.

Working alone or in pairs, use pattern blocks to create a unique design. Challenge yourself to think about what fractional part each block represents in relation to the whole (entire design).

As a group, debrief using the following discussion questions:

- How did you calculate the fractions for each pattern block piece?

- What other relationships did you discover while engaging in this task?

REFLECT

- How did patterns play a role in your experience?

- How might students approach this task?

- What other ideas do you have for using pattern blocks in your classroom?

Game Time

This month's games extend the work from Manipulatives and Models Matter and allow students to explore counting, decomposing shapes, equivalent shapes, and spatial awareness.

COVER THE SHAPE

Materials:

4 players

Pattern blocks

Cover the Shape pattern

Pattern Block Spinner

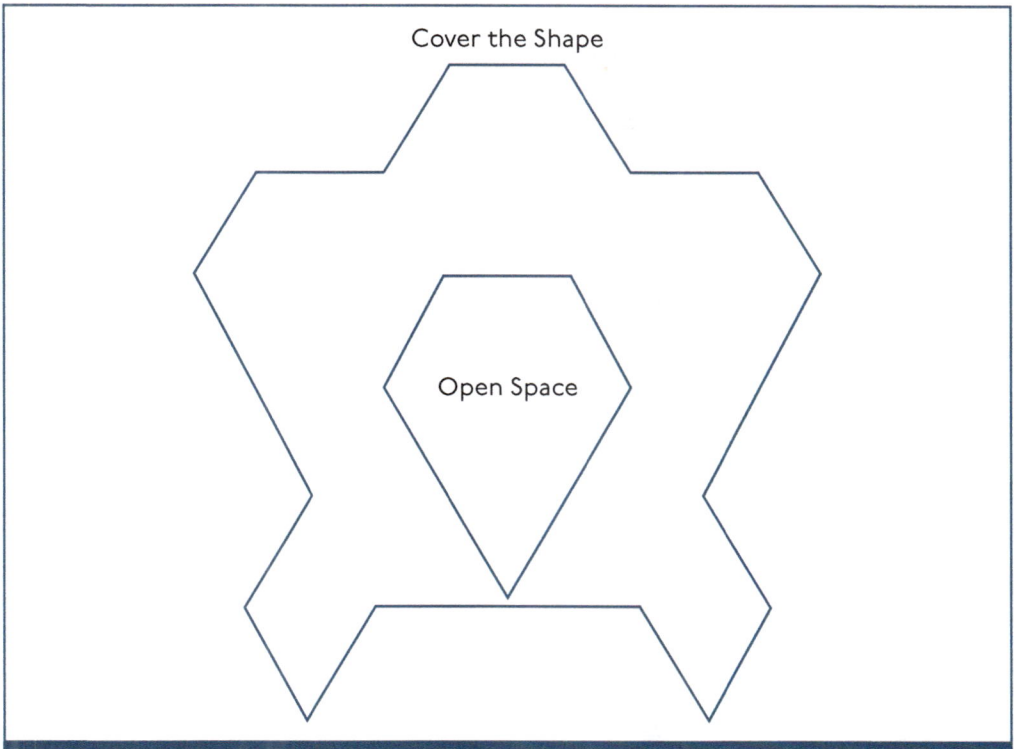

Cover the Shape

Open Space

Challenge: Be the first team to cover the game board in pattern blocks.

Game directions:

1. Each team of two lays out their own shape pattern in front of them.

2. On your team's turn, spin the Pattern Block Spinner. You have one of two choices:
   - Use the block that was shown on the spinner to place it on the game board.
   - Use smaller blocks that are equivalent to the block that was shown on the spinner and place them on the game board.

3. Blocks that have been placed cannot be moved.

4. If, at any time, a block cannot be placed on the board, that team loses its turn.

5. Team 2 now spins the Pattern Block Spinner and repeats Step 2.

6. The first team to cover their shape pattern is the winner.

Discuss the experience:

- How did you decide where to place your pattern blocks?

- How did your strategy change throughout the game?

(Continued)

(Continued)

- What math habits would students be using in playing *Cover the Shape*?

- How are patterns used during gameplay? How do they help the players?

COMPLETE THE SHAPE

Materials:

2 players

Pattern Blocks

Hexagon Pattern Game Board

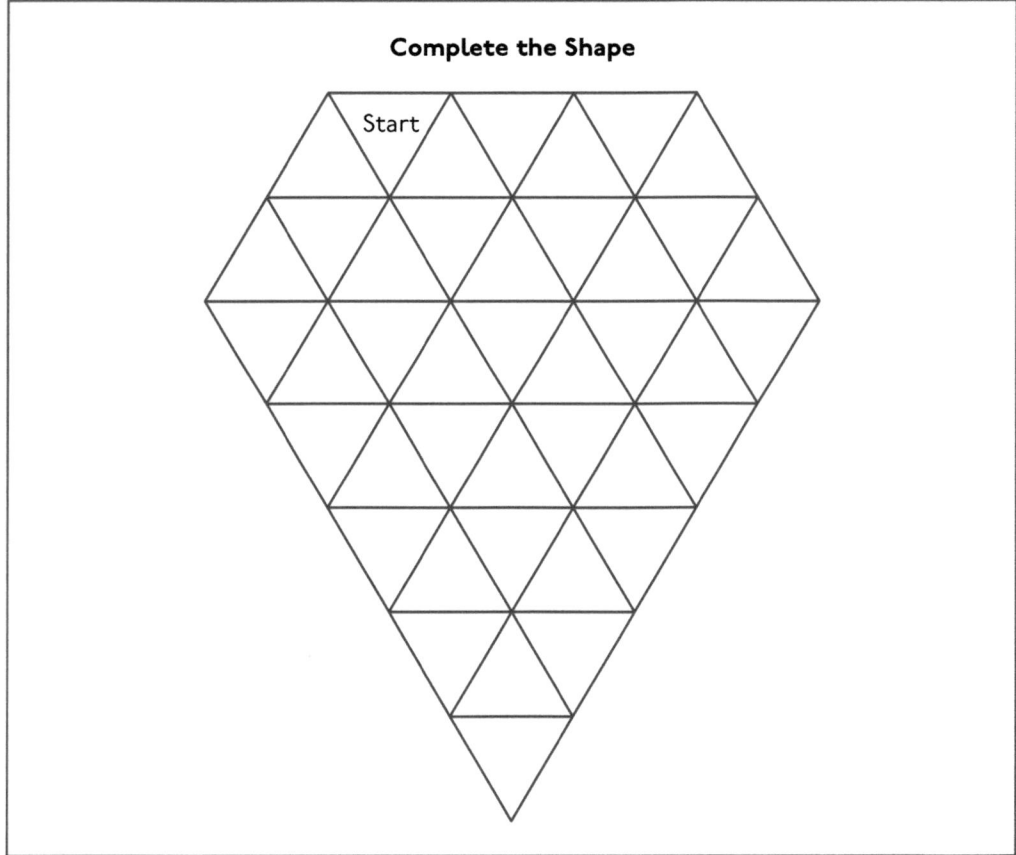

Complete the Shape

Challenge: Be the player to place the last pattern block on the game board.

Game directions:

I.   On your turn, begin by placing a green pattern block on the spot marked "start" on the game board.

2. Your opponent selects a pattern block to add to the board and satisfies the following:

    - One side of each new block must touch an existing block on the board.

    - The sides that touch must touch completely (touching corners is not allowed).

3. Continue playing back and forth by adding one block at a time. The player who places the last pattern block is the winner.

Discuss the experience:

- What strategies did you use to place your blocks?

- What math habits would students be using when playing *The Last Pattern Block*?

- How are patterns used during gameplay? How do they help the players?

 Available as a downloadable resource on the companion website.

IN YOUR CLASSROOM

Classroom Story

In Mr. Kapsalis's class, his fifth graders were beginning their study of expressions. He decided to use dot images to generate patterns in numbers as a jumping-off point for writing simple expressions. Students were shown the following image and asked to find as many ways as they could think of to calculate the total number of dots.

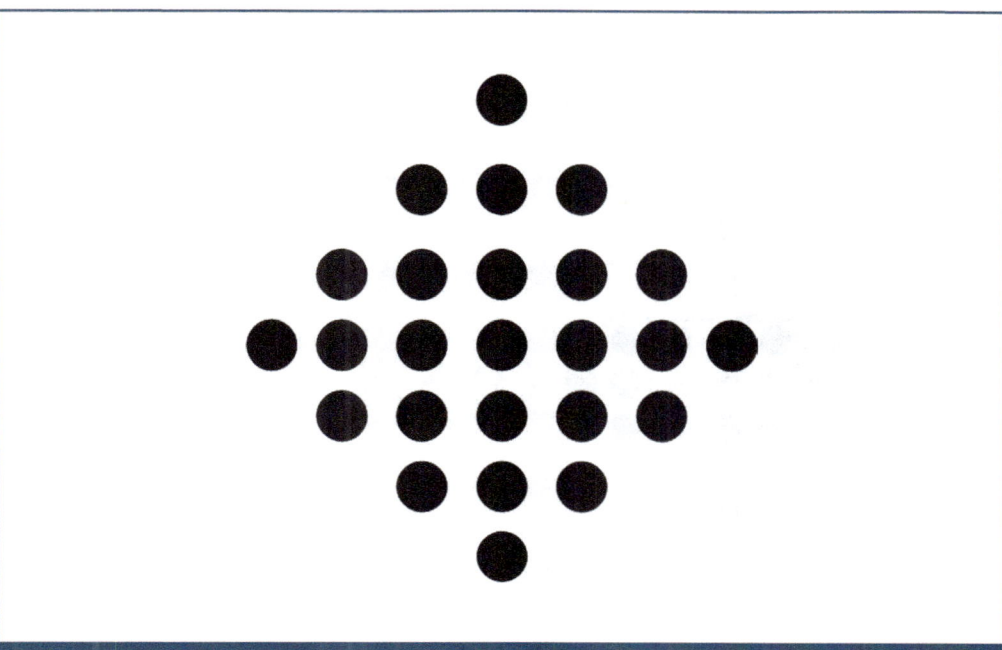

Leo: *I found a square in the middle and four groups of four around the outside.*

Mr. Kapsalis: *So you found a repeating pattern around the outside and a square pattern in the middle of the design. If we wanted to write a number sentence to describe each of those separate parts, what might we write?*

Leo: *Well I know that the middle would be 3 plus 3 plus 3.*

Mr. Kapsalis: *Is there another way to write that?*

Yuming: *We could write 3 times 3.*

Mr. Kapsalis: *Ok, so we have two ways to write the middle. What about the outside?*

Leo: *Well four groups of four would be 4 times 4. So, I have 3 times 3 and 4 times 4. I have 25 altogether.*

Mr. Kapsalis recorded the following equation next to the image to demonstrate each part of the pattern and its connection to the image.

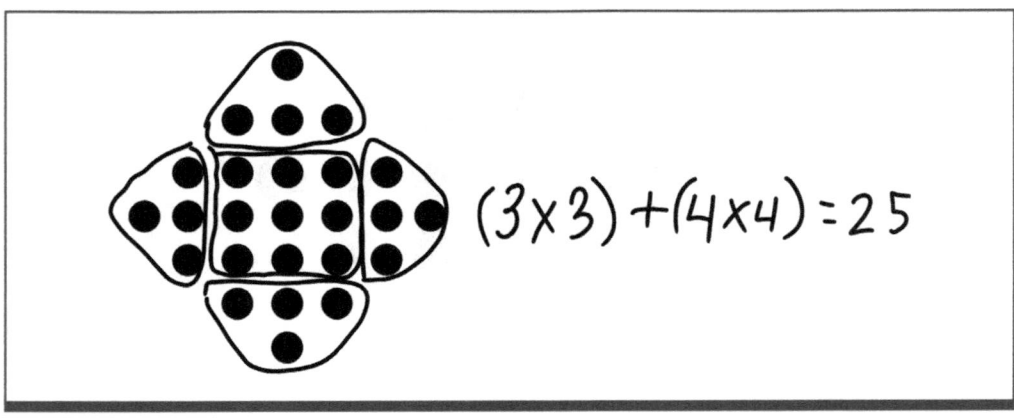

Mr. Kapsalis: *Did anyone see this another way?*

Marisol: *I noticed a pattern diagonally. Starting at the left, I saw a group of four, a group of three, and then back and forth between groups of four and three. There were four groups of four and three groups of three.*

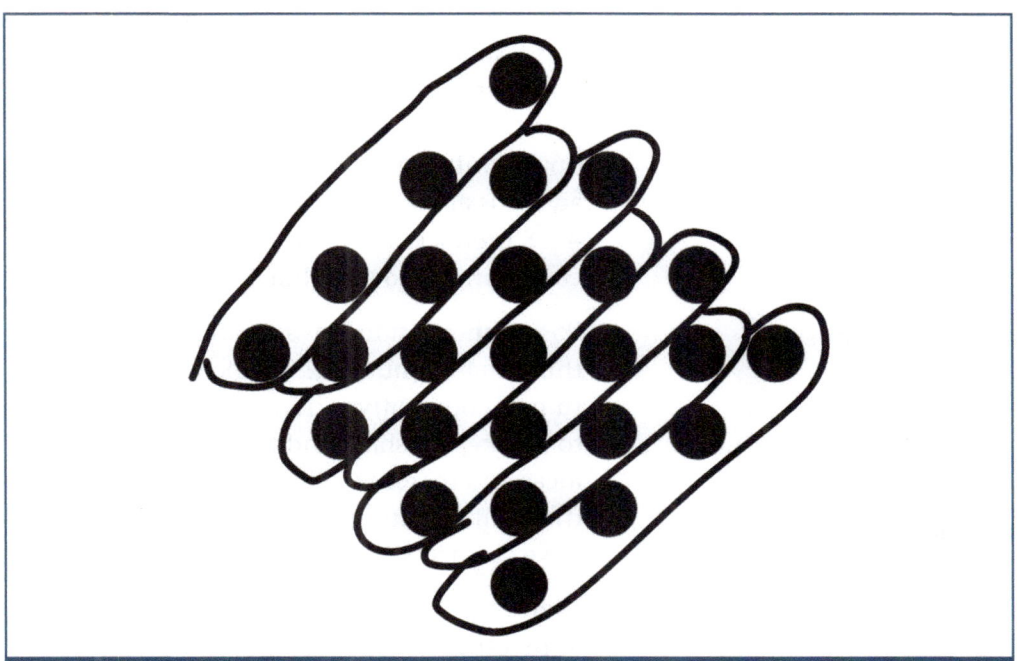

Mr. Kapsalis: *Just like Leo's, how can we quantify what Marisol saw?*

Antonello: *It's just like Leo's!*

Mr. Kapsalis: *How do you mean?*

Antonello: *She also has three groups of three or 3 times 3 and four groups of four or 4 times 4. You can write the same equation.*

Many other students nodded their heads in agreement. Mr. Kapsalis wrote the equation next to the image. He wanted to make the connection of the equation back to Marisol's visual of the dots. He decided to focus on the expressions within the equation and asked students to point out where the 3 times 3 was in her drawing. He then asked if they could see the 3 times 3 in Leo's visual.

A number of other students shared the ways in which they saw the arrangements of dots. Each time, Mr. Kapsalis would ask students to name the expressions and identify them within the dot images. This lesson would serve as an anchoring experience for students to think about creating and naming expressions. Mr. Kapsalis was satisfied by his students' ability to look for patterns and use those patterns to create numerical expressions.

Anchor Lesson

Learning to Notice Patterns

WHAT?

In this lesson, students will become pattern hunters as they explore consecutive sums. This engaging exploration can be found in Burns' (2020) publication *Welcome to Math Class*.

YOU NEED:

A series of related computational problems

Chart paper (2)

Markers

Hundreds Chart (optional)

HOW?

1. Write "We are pattern hunters" at the top of an anchor chart. Set this aside for the end of the lesson.

2. Prepare a chart with numbers 1–30 listed vertically. Leave room to the side so you can add student equations as they find them.

3. Display a hundreds chart for students to look at. Highlight a couple of different series of consecutive numbers. For example, 2, 3, 4 and 13, 14, 15 and 41, 42, 43. If you don't have a hundreds chart, write a series of consecutive numbers on the board for students to see.

4. Ask students to think about what they notice. If students don't mention the consecutive numbers, tell them the word and ask students if they can give you a few examples. You may even use a couple of counterexamples for clarity.

5. Pose a simpler problem. Could we write 9 as a sum of consecutive addends? Students may list any addends that give them the sum of 9 and use the opportunity to verify which sums of addends are consecutive. For example, 4 + 5 produces a sum of 9 using consecutive addends, but 2 + 7 does not.

6. Pose a question for investigation. Do you think that all numbers can be written as a sum of consecutive addends? Ask students to think–pair–share and listen to their ideas and reasoning. Tell students that they will be pattern hunters today to see if we can answer our question.

7. Divide up the numbers for groups to investigate. Students should work together to list which numbers can be written as a sum of consecutive addends. When they find one, they should record it on the class chart.

8. Give students time to work.

9. At the end of the work time, bring the students back together to discuss what they found and what they noticed. Students should refer to the class chart that was created during the work time. Use the following questions to guide the discussion:

   - What patterns are you noticing in our numbers?

   - How might you predict what other numbers can be written as a sum of consecutive addends?

   - If someone else were to encounter a problem like this, what would you tell them?

10. Students may not come up with a full explanation about which numbers can be written as a sum of consecutive addends during this experience, but you may consider returning to the investigation as time allows.

11. Show students the anchor chart "We are pattern hunters" and ask them to consider how they hunt for patterns. Record students' ideas under the title. Here are some ideas to get you started:

    - Write down a lot of related problems and stop to look at them.

    - Ask questions.

    - Look for things that keep happening.

    - Ask yourself why something happens or doesn't happen.

REFLECT

The habit of finding, using, and applying patterns lies at the heart of mathematics. Our students need ample time to sit with math problems and think and speak about what looks the same, using the connections between mathematical concepts and problems.

Use the following discussion questions to reflect with students:

- How did you use a pattern to help you solve the problems?

- Why might it be useful to seek out patterns in math?

- If we slow down and notice patterns within problems, how might that help us as math learners?

LET'S DO SOME MATH ACROSS OUR SCHOOL!

Share a photo of a visually interesting pattern of objects. You might use one of Howie Hua's photos from the link that follows or take a photo yourself of everyday objects students are familiar with. You can also encourage students to notice patterns when they are outside of school. They could use a parent's cell phone to take a picture of a pattern they see and send it to you to be displayed for the class to see, count, and talk about patterns.

Two examples are shown here.

How many triangles do you see? How did you see them?

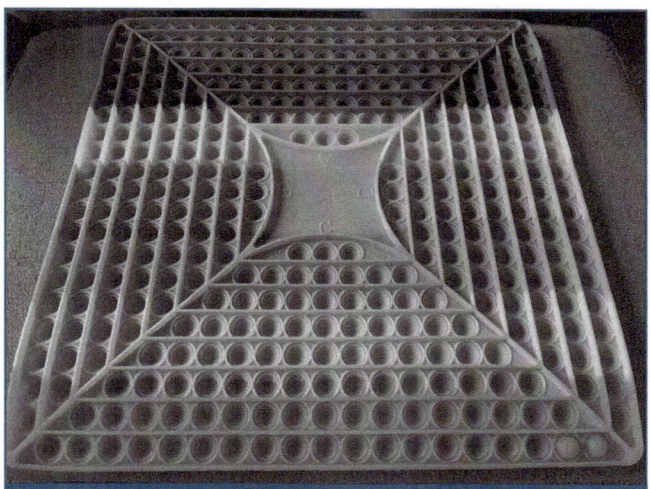

How many circles do you see? How did you count them?

Ask students to think about how many objects they see and how they see them. Record their strategies visually and with expressions and equations as in the vignette above. Use the Talk Facilitation Moves of eliciting, clarifying, probing, and orienting, and encourage students to use the Talking-to-Learn Moves of explaining and justifying, connecting, conjecturing, critiquing, and questioning from Chapter 7—January to discuss the strategies shared. (Hua, 2023, January 15; https://bit.ly/3P9FEE0)

TEACHING MOVE: PATTERNS TO GENERALIZATIONS

Once students begin to recognize patterns in numbers, problems, content, or even representations, they are ready to begin generalizing from these patterns to mathematical rules that help them solve future problems.

For example, a kindergarten student might notice that when they use counters to find the pairs of addends with a sum of five, one of the addends goes up, while the other addend goes down.

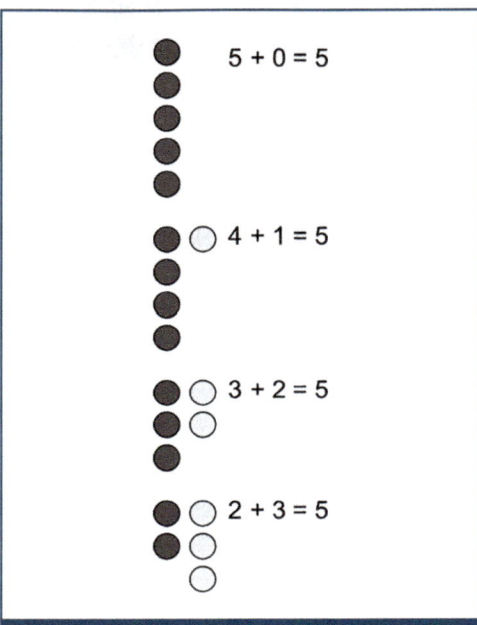

We can ask students, "Do you think this pattern will work with other numbers?" This question encourages students to consider big ideas in our number system such as the counting principle of inclusion, the commutative property, and even the operation of adding. Students can then test out their theories and report back on their findings.

REFLECT

- How might asking students to solve and think about related problems help them develop the habits of a mathematician?

- What is a content example from your grade level in which students could use patterns to formulate generalizations?

CLASSROOM ROUTINE: PATTERNS IN CLASS DATA

We can continue building our powerful math communities by seeking to know more about the students in our classroom and using the information to have mathematical conversations.

Once a week, post a question that students can answer about themselves. Have students write their responses on a sticky note and create a visual graph or have them add their name to a premade graph.

For example: How many siblings do you have? Post your response by marking an "x" in the correct column.

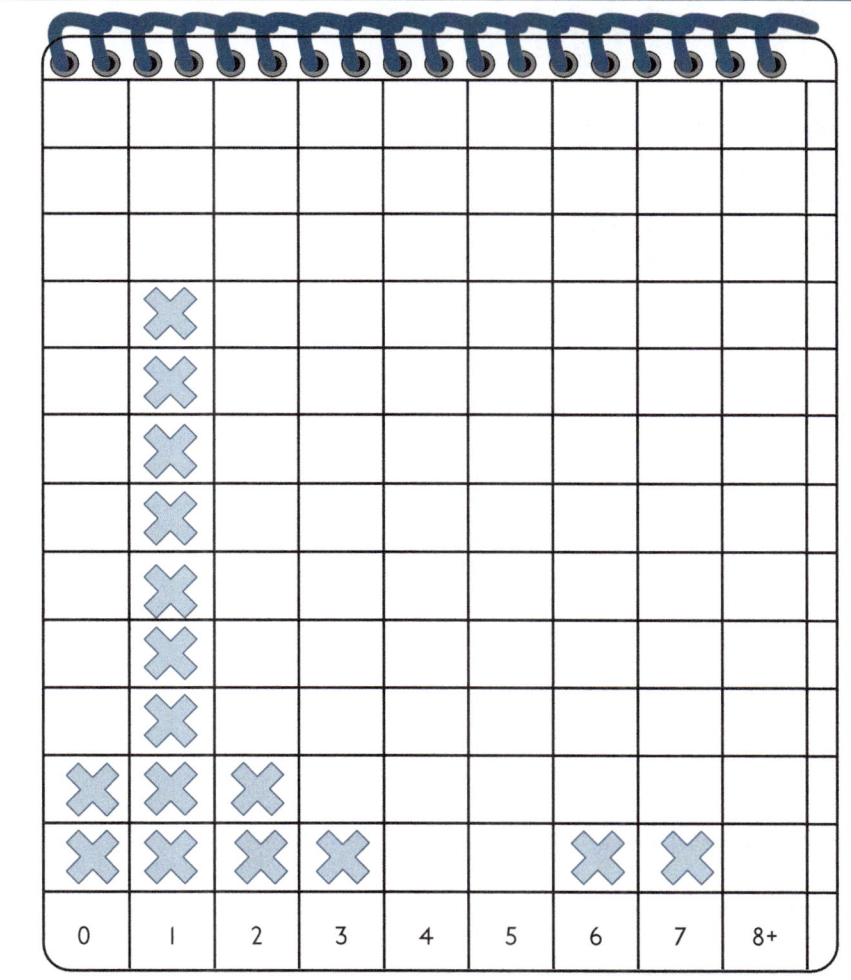

0	1	2	3	4	5	6	7	8+

(Continued)

(Continued)

Discuss the results with your class using questions like these:

- What patterns do you notice in our classroom data?

- Do you think this data would be different for other classes? Why or why not?

- What math questions could we ask that can be answered with this data?

Be sure to select questions that allow all students to participate in the data collection and analysis. Questions like "How many presents did you get for Christmas?" or "How many bedrooms are in your house?" can inadvertently cause students to feel that they don't belong or are less valued members of the class community.

STATION ACTIVITY: PROBLEM STRINGS

Student learning is maximized when students have the opportunity to revisit, discuss, analyze, and apply mathematical ideas over time. In this month's anchor lesson, students used patterns to analyze and discuss commonalities across a string of related computational problems.

Consider statements students might make about this string of subtraction problems:

- $60 - 15 =$

- $50 - 15 =$

- $40 - 15 =$

- $30 - 15 =$

Statements students might make about the problems include:

- You can subtract ten and then five more.

- When you subtract a number with five in the one's place, the answer is a number with a five in the one's place.

- You can count on from 15 using the hundreds chart to get the answer.

For this month's station activity, students will use a string of numbers to create statements about patterns they notice. The goal is to have students practice looking for patterns, naming those patterns, and seeing if they hold true for other numbers.

Select a string of problems for students to explore at the station activity. For their time at the station, students should make a statement about a pattern they notice in the problems and try that statement out on another set of numbers.

Students should keep their work to debrief as a class at the end of the week.

An example of the station activity follows.

Statement: When you subtract 15, you can subtract 10 and then 5 more.

Other Examples:

Does your statement still seem to be true?

Did you find examples where it doesn't work?

For younger students, consider having them generate more of the same pattern from Teaching Move: Patterns to Generalizations on p. 344 to bring to the class at the end of the week.

For example, you may have students explore decomposing a different number. Students may be given a handful of two-color counters of a number of their choosing, shake and

(Continued)

(Continued)

spill the counters, and record the different ways that number can be shown with two addends. At the end of the week, students could bring their station worksheet with them as you discuss the station activity.

Number I explored:

6

My Picture:

My Equation

4 + 2 = 6

Debrief with the class at the end of the week. Use the following reflection questions to guide the discussion:

- What were some of the other numbers you explored in the station activity? (record student ideas)

- Did you discover any other patterns? What were they?

LITERATURE CONNECTIONS

Pitter Pattern **by Hess Elbert** (2020) (Grades K–2)

Across a week, Lu and her friends find repeating and growing patterns in common objects, in nature, in music and movement, and in time. In the end notes of this engaging book, the author says, "A pattern is a design that's predictable. It's often something that repeats. We use patterns to help predict what comes next and to help us understand how the world works" (Endnotes, para. 1).

Activity: What kinds of patterns can you find in your daily activities? How might you represent these patterns with mathematics?

Seeing Symmetry **by Leedey** (2012) (Grades 3–5)

Using a series of fascinating illustrations, Loreen Leedy introduces the concepts of line symmetry and rotational symmetry. She encourages readers to look for symmetry in nature, in letters and words, in the design of machines, art, architecture, furniture, and in our own bodies and clothing. She discusses how symmetry relates to mathematics concepts, including patterns, equivalence, and transformations.

Activity: Loreen Leedy says that line symmetry allows many animals, including human beings, to move and that rotational symmetry allows many machines, such as helicopters, to function. What are some other examples of mathematical patterns that are not only interesting and beautiful to look at but also useful? What is a mathematical pattern that is important to you, and why?

SPOTLIGHT ON BRAIN SCIENCE

The human brain, like the brains of all other mammals, has a unique layer called the **neocortex** composed of numerous gray folds with millions of pattern-recognizing cells. The neocortex of our brains craves patterns and uses the patterns it finds to make connections that support memory and learning (Bor, 2012). Since our brains are wired to recognize and use patterns, we can help students to know that we should seek out, replicate, and analyze patterns in mathematics.

You can share the following information with students in a mini lesson. You might also send this reproducible home along with the Family Newsletter to help families talk with their child about these important ideas.

OUR BRAINS LOOK FOR PATTERNS.

Human brains are made to look for patterns. A special layer in the brain called the **neocortex** has millions of pattern-recognizing cells. When you look at a picture of the brain, the neocortex is the largest part of the brain with all of the folds.

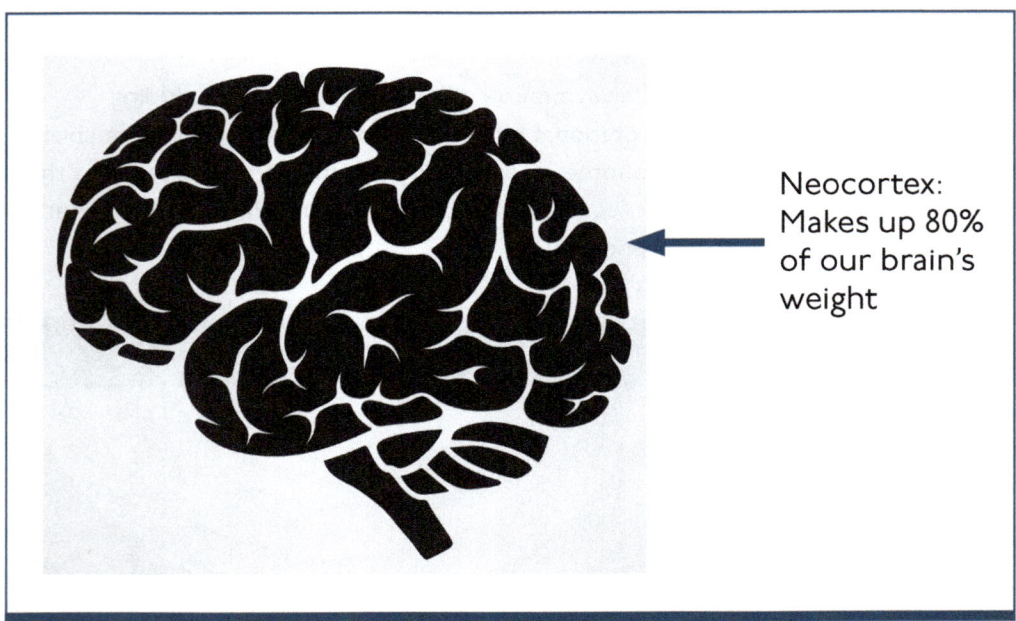

Neocortex: Makes up 80% of our brain's weight

Image source: istock.com/Nazarii

Our brains are always looking for patterns. Looking for and using patterns can help us to learn math. Here are some things you can do:

- When you look at a new math problem, stop and think, "Does this look like anything I have seen before?"

- When you compare two math problems or representations, stop and notice, "What is that same? Is there a pattern?"

- Speak up when you notice a pattern. Share your noticing with others. It probably means you have found a big idea.

Choose one of the following ways to think more about this important information:

1. Write down why these ideas are important to you and how you want to put them to work in your life.

2. Talk to a friend about these ideas. Decide together how you will put them to work.

3. Share these ideas with someone at home. Tell them why these ideas are important to you and how you plan to use them.

 Available as a downloadable resource on the companion website.

SPOTLIGHT ON EQUITY: PATTERNS THAT VALUE DIVERSITY

Different cultures have unique patterns that are meaningful to them. For example, Indigenous cultures have a deep relationship with the cosmos, and the Star Quilt blends a star design with the highly skilled art of crafting blankets. Although the earliest blankets were made from resources found in their respective locations such as animal hides and sinew, modern Indigenous Star Quilts are made from beautiful fabrics with a recognizable eight-point star sewn on a quilt and given to a loved one during a time of grief or loss. It signifies the loved one's journey through life.

TRY IT

We can honor and respect other cultural groups by recognizing some common patterns and their meanings. Bringing in community members who can share their knowledge of unique patterns in cultures is a great way to expose students to these ideas. You might seek out opportunities to take students to see artwork in a gallery that depicts different cultural patterns.

Within the classroom walls, consider doing a mini study on patterns in culture. Show students examples of what the patterns look like and discuss the meaning behind them. You may even show students a map that connects the country of origin while recognizing the people who share the customs and cultures that live in and around us. Use examples that may be historical but also modern for students to connect to their contemporary world.

Seeking out museum websites that offer a history and images of unique cultures is another great way to accurately depict mathematical patterns in art and culture.

Consider checking out these global patterns from an artist named Michelle Carlos (Carlos, 2020; https://bitly/3lpQ7ar).

MATHEMATICAL ME: STUDENT JOURNAL AND PORTFOLIO

Student Journals:

Have students respond to the following questions:

- How does looking for math patterns help me learn?

- How can I use math patterns to solve new problems?

Older students can write their responses in their Mathematical Me journals. Younger students can draw a picture showing their response, or you might gather responses during a class discussion or quick one-on-one interviews.

Review students' responses to monitor students' beliefs about mathematical patterns. You might summarize this data by tallying or graphing the different ideas students offer.

Be sure to share and discuss this data with your class. Ask, "What does this data tell us?" and "What are some things we can do to continue growing as mathematicians?"

Student Portfolios:

Have students choose a piece of work where they identified or used a pattern. On a sticky note, have them write the date and complete this sentence frame:

Source: Istock.com/Thammask-Chuenchom

FAMILY NEWSLETTER: WE SEE AND USE PATTERNS!

WHAT WE'RE LEARNING

This month we're learning about how mathematicians seek out and create patterns to think about math. Patterns are useful for learning math because students can look at patterns to identify and name math concepts. For example, students look at the following multiplication facts and see a pattern in the products. They might say that any time a number is multiplied by two that number doubles.

$2 \times 2 = 4$

$3 \times 2 = 6$

$4 \times 2 = 8$

$5 \times 2 = 10$

$10 \times 2 = 20$

WHY IT'S IMPORTANT

Math is all about patterns. There are patterns in numbers and shapes, and there are patterns in the world all around us. When we see a math pattern, it's important to notice it and think about what it means. This is how we discover new mathematical ideas and connect our old thinking to new thinking.

TRY IT OUT

What patterns do you see in these numbers? How could patterns like these help your child learn math?

- 0.01

- 0.1

- 1

- 10

- 100

(Continued)

(Continued)

HOW YOU CAN HELP

As you move about your day, talk with your child about patterns that you notice. Here are some ideas to get started:

- I see that every license plate has a number and a letter. See if you notice it too.

- There seem to be a couple of street lights on every block. Do you think they are all spaced the same distance apart?

- Our dozen eggs are placed in two rows in the egg carton. If we had a different-sized carton, what are some different ways we could arrange the eggs?

Notice mathematical patterns in nature. What is interesting to you? Share these patterns with your child. Help them to become an avid noticer of patterns.

Pay attention to interesting visual patterns. Take a picture with your phone and show your child. Be curious. Ask them to tell you the pattern they see or have them count how many objects are shown.

Many adults and students believe that math is only found in the math classroom at school. Help your child to see the math that is everywhere by looking at and finding patterns.

 Available as a downloadable resource on the companion website.

CHECKING IN ON OUR LEARNING: WE LOOK FOR AND USE PATTERNS!

- How does looking for math patterns help me learn?
- How can I use math patterns to solve new problems?

This month's learning focus on patterns is at the heart of doing math. Patterns live outside the classroom in the world around us, but they also help mathematicians learn about and discover math that is new to them. In a powerful math community, leaders, teachers, and students take time to notice and use patterns.

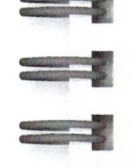

MATHEMATICAL ME: EDUCATOR JOURNAL

- How has this month's learning focus supported your students' mathematical growth?

- How has this month's learning focus supported your growth as a math teacher?

- How has this month's learning focus supported your school's growth as a powerful math community?

LOOKING AHEAD

In May, we will take the time to reflect on and celebrate our yearlong math learning journey. We will help students identify their mathematical strengths and set goals for their next steps as math learners.

Questions for Reflection

- What have I learned this year that empowers me as a mathematician and math learner?

- What have I learned this year that empowers me as a math teacher?

CHAPTER ELEVEN

May
We Are Powerful Mathematicians!

ESSENTIAL QUESTIONS

- What are my math strengths?
- How can I use math to do important things in and out of school?
- What are we celebrating about our powerful math community?

THIS MONTH'S FOCUS

In a powerful math community, students and educators know that their mathematical habits and understandings empower them to do things they care about in the real world, as well as in school. They can identify their current mathematical strengths and goals for future math learning.

Educators proudly acknowledge their role in helping students see themselves as mathematically powerful and experience mathematics as a useful and joyful pursuit. They recognize and celebrate each student's unique mathematical brilliance and growth.

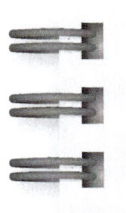

MATHEMATICAL ME: EDUCATOR JOURNAL

Take five minutes to respond to the essential questions. You will have a chance to reflect on these same questions and to have students respond to these questions at the end of the month.

ON YOUR OWN

LET'S DO SOME MATH!
MATH IN MY LIFE

Image source: istock.com/leremy

Think of something you are passionate about. It might be a hobby or an interest you pursue in your free time. Perhaps it's an important goal that you have for yourself. Or maybe it's a cause or a real-world issue you care deeply about.

Jot down some thoughts about this passion using the following sentence frames:

- My personal passion is_____.

- This passion is important to me because_____.

Now mathematize your passion. You might Google connections between your passion and mathematics to spark thinking. Complete the following sentence frames:

- One way math is connected to my passion is_____.

- One way math can deepen my understanding of my passion is_____.

- One way math can help me use my passion to make the world a better place is
  _____.

Look over what you've written. Now complete one final sentence frame:

I am a powerful mathematician because_____.

Save this sentence to share with your colleagues.

online resources   Available as a downloadable resource on the companion website.

Why This Focus

I am a powerful mathematician! As our students walk out the classroom door at the end of the school year, we want each of them to fully own this belief.

We are powerful mathematicians!

Image source: Istock.com/syntika

This belief statement embraces two important elements of a healthy mathematical mindset. When our students declare "I am a mathematician!" they claim their mathematical identity. They see themselves as a person capable of understanding mathematical ideas, a person who chooses to engage in mathematical activity on a regular basis.

When our students add the adjective "powerful" to their belief statement, they define mathematics as a tool that allows them to do important things in their lives, both in and out of school. They are asserting their mathematical agency.

Both aspects of a mathematical mindset are important. Students with strong math identities and robust mathematical agency will thrive in school and in life. They are also positioned to use math to make positive contributions to their communities, society, and the world.

How have your students grown their mathematical powers this year?

Imagine that you're looking through a window into your classroom. Observe your students engaged in mathematical learning work. What mathematical powers do you see in individual students and your class collectively? What specific ways have your students grown more mathematically powerful this year? What experiences have contributed to that growth?

How have you strengthened your own mathematical powers this year?

Imagine looking in a mirror at your own growth as a mathematician and math teacher. What are some ways that your mathematical powers have increased? What experiences contributed to that growth?

As you finish the school year, how will you measure your success as a math teacher? How will your students' and your own growing mathematical powers factor into your self-assessment of your accomplishments this year?

We choose to be educators to make a positive difference in students' lives. Perhaps we use test scores and other numerical measures of learning to verify that we are indeed making a difference, but in our hearts, we know this data doesn't begin to capture the most important aspects of learning. We want to send our students off with the mindsets, skills, and understandings they need for future success in school and life. We want them to be mathematically powerful.

> "Math class isn't just where we teach content to students—it's a story-making machine. It is where a math story is authored."
>
> —Orton (2022, p. 182)

> "Student agency is the goal, not test scores."
>
> —Safir and Dugan (2021, p. 4)

Orton (2022) encouraged us to "think of ourselves less as people who teach mathematics content to students and more as cultural stewards who help our students author their own math stories" (p. 154). He said, "A math story describes a person's relationship with mathematics. It is how they describe their math identity and how they exercise their sense of mathematical agency" (p. 14).

This chapter is about mathematical power. It is about the unique and important stories of the students in your class and how they have become more mathematically powerful this year. As you wave goodbye to your students on the last day of school, you send them on their way to continue their math learning stories, equipped with the mathematical powers you've helped them to develop this year. In this chapter, we'll look at how you can bestow a parting gift on your students—a keen awareness of the mathematical powers they possess.

Sue's Math Story:

Math helps me to be more effective in my work as a teacher educator and professional learning consultant. Math supports my core values of learning and service.

My professional passion is supporting and advocating for teachers. I care about helping preservice teachers see teaching as an important and satisfying career and about equipping future teachers with the understandings, skills, and mindsets they need to succeed in the classroom. I also care about helping inservice teachers remember their whys, their personal reasons for choosing to teach, and to see teaching as a learning profession that carries both the opportunity and the obligation to build their craft every day as they learn from and alongside their students. I believe that teaching and learning can be powerful vehicles for making the world a better place.

I am continually applying mathematical ideas as I engage with my passion. I pay careful attention to the number of students in my math methods courses who pass their teacher exams, complete their teacher education program, and land their first teaching jobs. When I facilitate professional learning sessions for in-service teachers, I review the session feedback data as a measure of the learning value of the session for teachers. I also find myself thinking about the number of students each of these teachers will impact across their careers.

Math is a tool I use to deepen my understandings related to my passion for teaching and teachers. I regularly study Hattie et al.'s (2017) meta-research about teaching practices that have the greatest impact on student learning. Math is also a tool I can use to address problems I uncover. I carefully consider the data about the reasons teachers who work in challenging contexts choose to stay in the classroom, thinking about how I can use these insights in my professional learning work to help teachers become more resilient.

I am a powerful mathematician because I use math to grow and to make a positive difference in my world.

Holly's Math Story:

Math helps me advocate for mathematical experiences in school to be joyful, meaningful, and positive for all leaders, teachers, and students.

I am passionate about changing the narrative for what it means to learn and do math. It is important to me that leaders and teachers continue to remain curious about their mathematical skills and active in seeking meaningful learning opportunities for themselves and their students. I want my community to engage with mathematics in a positive way and see how mathematics makes the community and its members powerful.

When I work with any group of teachers and leaders, I make doing math together a priority. We explore the fun and beauty that math has to offer and practice math habits that bring everyone into the learning. We communicate, celebrate, laugh, analyze, question, and reflect. I intentionally set up my professional learning spaces to be places where teachers and leaders can explore their own mathematical identities.

Math has given me the confidence to tackle any challenge. I feel powerful knowing that mistakes, challenges, and flexibility are necessary elements of learning and doing math. I no longer feel limited; instead, I relish solving new problems. This freedom carries over into my life outside of mathematics, where I know that I can successfully navigate new challenges.

I am a powerful mathematician because I use math to bring joy to people's lives.

Mathematical Points of Power

Kobett and Karp (2020) defined a **mathematical point of power (POP)** as something "students do well in mathematics, which they not only state but rely upon when they face a novel challenge or problem" (p. 176). They advised teachers to "seek out and acknowledge these POPs while giving students opportunities to both practice and showcase their POPs" (p. 176). You have been doing precisely this all year!

We all have strengths related to mathematics content, as well as to mathematical practices or processes, the habits of mathematical thinking. We encourage you to help students recognize both types of strengths, but this chapter will focus primarily on developing your own and your students' awareness of their strengths with the math habits.

Tapping Into Your Experience

A first step in helping our students recognize their mathematical strengths or points of power is to take an inventory of our own mathematical strengths.

Step 1: Rank order your confidence using the 10 habits of mathematically powerful people. Mark the habit that you are most confident about with a 1, the habit that you are next most confident about with a 2, and so on.

_____ I expect math to make sense.

_____ I love challenging math problems.

_____ I learn from math mistakes.

_____ I see myself as a mathematician.

_____ I talk about math.

_____ I represent math in different ways.

_____ I make math connections.

_____ I look for and use patterns.

_____ I enjoy engaging in math, see math as a tool to do important things in real-world contexts, and as a result, am always becoming more mathematically powerful.

_____ I recognize that others are also mathematically powerful and that communities of mathematicians are powerful sources of support and learning.

What do you notice? What do you wonder?

What are the implications of this data for your teaching? For your use of mathematics in real life?

Step 2: For each of the habits of mathematically powerful people, how confident are you in your ability to develop this habit in students?

	NOT AT ALL CONFIDENT 1	A LITTLE CONFIDENT 2	CONFIDENT 3	QUITE CONFIDENT 4	HIGHLY CONFIDENT 5
Mathematically powerful people expect math to make sense.					
Mathematically powerful people love challenging math problems.					
Mathematically powerful people learn from math mistakes.					
Mathematically powerful people see themselves as mathematicians.					
Mathematically powerful people talk about math.					
Mathematically powerful people represent math in different ways.					
Mathematically powerful people make math connections.					
Mathematically powerful people look for and use patterns.					
Mathematically powerful people enjoy engaging in math, see math as a tool to do important things in real-world contexts, and, as a result, are always becoming more mathematically powerful.					
Mathematically powerful people recognize that others are also mathematically powerful and that communities of mathematicians are powerful sources of support and learning.					

What do you notice? What do you wonder?

What are the implications of this data for your students' learning? For your students' use of math in real life?

Cultural Wealth and Funds of Knowledge

Closely related to the idea of mathematical strengths is the concept of cultural wealth or funds of knowledge. All people possess a wealth of knowledge and skills acquired through life experiences and their families' and communities' unique cultures. Our students' funds of knowledge and cultural wealth are an asset to our students personally and to our classroom and school learning communities. We are all enriched by others' ways of thinking, and we are collectively more powerful when we connect with and learn from people whose experiences and perspectives differ from our own.

To be mathematically powerful, people need to recognize the utility, joy, and beauty inherent in mathematics. Therefore, students' mathematical experiences in and out of school must intersect with who they are as human beings. This intersection encompasses our students' cultural wealth and funds of knowledge. Kobett and Karp explained:

> We bring the child into the mathematics by bringing the mathematics into the child's life. . . . We use students' names, the names of streets near where they live, places in their neighborhoods, and the familiar things, people, and experiences in their world as components of the tasks we present. Our attempts are not merely to capture their world but also to make the connection to a future life that is filled with problem solving related to mathematical situations a REALITY. (2020, p. 147)

Students' relationships with mathematics become a strength when students see how mathematical activity connects with their lives (see Figure 11.1).

Figure 11.1 • Students Have a Strong Relationship With Mathematics When They See How It Connects With Their Lives

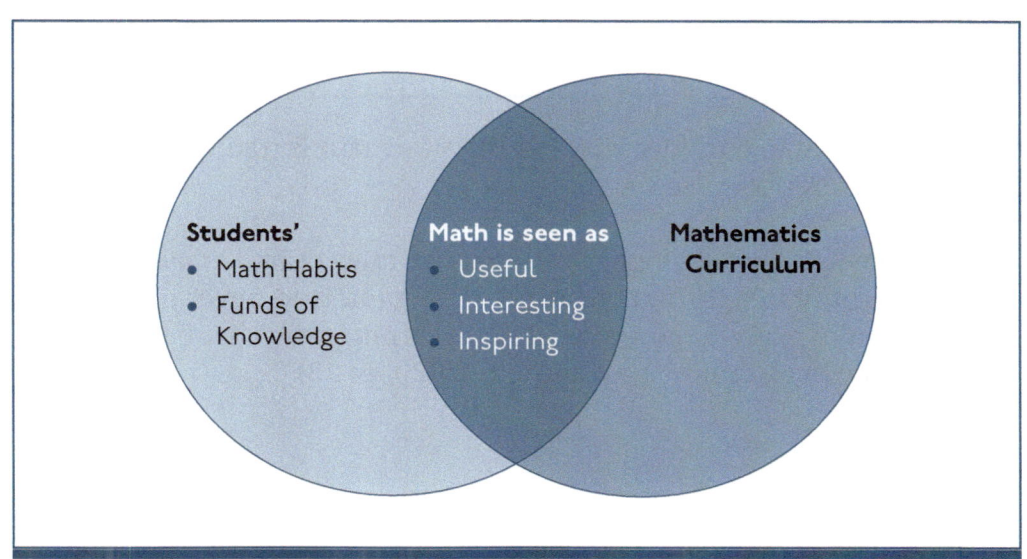

"All children should have the opportunity to experience the wonder, joy, and beauty of mathematics, and they should leave elementary school more curious than when they entered and see mathematics as even more wonderful, joyous, and beautiful."

—National Council of Teachers of Mathematics (NCTM, 2020, p. 45)

Math Joy

When I (Sue) self-assess my success as a math teacher of children and adults, one of the measures I use is evidence that my students regularly experience joy in their mathematical learning work. Here's what the NCTM has to say about joy and math:

> Joy occurs through making meaning, understanding, and seeing connections across mathematical representations. Children experience joy when they are encouraged to be creative and are provided with mathematical choices, such as generating their own approaches and strategies for solving mathematical problems. (2020, pp. 20–21)

> Joy occurs while being in the moment with mathematics, being proud of your accomplishments, and feeling the intrinsic reward of persevering through the productive struggle associated with solving mathematical problems. This joyfulness occurs most often in child-centered classrooms that promote active, inquiry-based, language-rich exploration of important mathematics. These experiences can support children's positive mathematical identity. However, for children to experience the wonder and joy in learning mathematics so must their teachers. (2020, p. 21)

Think of a moment this year when you experienced joy in teaching math.

Think of a moment when you experienced joy as a math learner.

What do you see and hear that lets you know students are experiencing math as joyful?

Why is it important for teachers and students to find mathematical activity joyful?

Problem Posing

Su challenged us to consider whether the way math is traditionally taught in school makes sense. He stated:

> Picture what it would be like to learn the rules of basketball and practice only free throws but never see a game and never play – until you're ready to go professional. Learning wouldn't be joyful, and you wouldn't now be prepared. (2020, p. 23)

Su encouraged math educators to give students authentic experiences engaging in real mathematics: "School mathematics sets you up for future exploration, but imagine how different our experiences would be if we could explore math now, as we learn it" (2020, p. 23). This genuine mathematical exploration, Su said, begins with the questions we naturally pose as we navigate aspects of life that are important to us.

As I (Sue) think about my passion for math education and teacher advocacy, multiple questions involving mathematics come to mind:

- Knowing that mathematical identity and agency are the key to mathematical proficiency, how can we quantify and measure growth in these critical aspects of learning?

- How can we gather classroom data that reveals our unconscious biases and inequities in classroom practices to create mathematics learning spaces that provide all students with the support they need to flourish?

- Research has suggested that teachers need as many as 50 hours of professional learning and practice to become proficient with new instructional practices (Aguilar & Cohen, 2020, p. 130). Knowing that teachers, like all learners, need time and support to internalize new skills, how can we weave this needed learning time into the collaborative structures of school life?

Go back to the notes you took earlier about your personal passion. Brainstorm three questions you might ask related to this passion that could be answered using mathematics. Jot one of your questions down to share with colleagues.

Posing mathematical questions related to things we care about is an important way to develop our mathematical eyesight and to strengthen our math identities and agency. If we are committed to helping our students see themselves as mathematicians and to developing their mathematical powers, then we need to first cultivate our own ability to pose mathematical questions that are personally meaningful. Then we will be ready to integrate opportunities for students to pose mathematical questions that connect to their lives into our math instruction. You will find ideas for ways to use problem posing throughout this chapter.

> "When children are given opportunities to consider how the mathematics they are learning in school relates to them, as well as pose and explore their own mathematics problems and ask their own questions, they become more deeply engaged in the mathematics and perform at higher levels."
>
> —NCTM (2020, p. 47)

TOGETHER WITH YOUR TEACHING COMMUNITY

Since We Met Last (10 minutes)

In April, you and your colleagues collected some classroom data about patterns in your curriculum and in real life.

Use the following protocol to share and reflect on this data with your math teaching community:

- Share your data with a colleague. (5 minutes)

- As a group, discuss what you learned from the data and how these insights can help you to support your students' growth as mathematicians. (5 minutes)

Let's Do Some Math Together! (10 minutes)

Share your "I am a powerful mathematician because…" statement and the mathematical question you generated about your passion with your team.

Discuss:

- How has your definition of mathematics and your relationship with mathematics changed as a result of your experiences this year?

- How do you see your students' definitions of and relationships with mathematics growing? How will this impact their future relationships with math?

Building Our Expertise (30 minutes)

On a daily basis, we help our students understand new math concepts and build proficiency with mathematical skills. Although this aspect of our math instruction is critical, we also need to periodically give students opportunities to use math to understand and solve real-world problems.

Hattie et al. (2017) called this essential aspect of our curriculum "transfer learning." They explained, "Learning demands that students be able to apply—or transfer—their knowledge, skills, and strategies to new tasks and new situations." Transfer learning allows students to "take the reins of their own learning, think metacognitively, and apply what they know to a variety of real-world contexts" (p. 32). Until students can transfer the skills and understandings they have learned to new contexts, their learning is not complete.

McHugh (2023) echoed this important message: "Students deserve to experience how mathematics can be used to authentically engage with their world" (p. 27). She told us, "When students engage in a mathematics project by asking questions and constructing answers, they recognize that mathematics does not just live in a textbook, but rather can be used as a powerful tool in our society" (p. 73).

You and your team will work together to plan an end-of-year math investigation to engage your students in synthesizing and applying their math learning within a meaningful real-world context. This investigation is designed to strengthen students' math identities and agency and remind students of the importance of the mathematics they've learned.

The investigation will follow a **Here's what! So what? Now what?** structure:

- **Here's what!** Students consider a real-world topic or issue that involves mathematical ideas.

- **So what?** Students use mathematics to study this topic or issue.

- **Now what?** Students reflect on what they learn from their investigation and commit to an action step in response to their insights.

(Continued)

(Continued)

Teacher Preparation for the Math Investigation

STEP 1: IDENTIFY A TOPIC OR ISSUE AND BRAINSTORM QUESTIONS.

Together with your team identify a topic or an issue that is relevant and interesting to your students. It could be a topic the class has studied in another subject area (e.g., an environmental issue discussed during Earth Day) or an issue significant to the school community (e.g., food insecurity or access to clean water). Brainstorm questions related to this topic that could be answered using mathematics.

Example: Students' access to books and support for reading over the summer (Note: This example was chosen because of its potential to encourage summer reading):

- *What hours are local libraries open this summer?*

- *Can families travel conveniently and safely to these locations during hours when these libraries are open?*

- *How many books can be checked out at a time? What is the checkout period for books?*

- *How long does it take to read a book? If a family visits the library once a week, how many books should they check out each visit?*

- *How much does an average book weigh? If a family checks out multiple books for every family member, how much will these books weigh altogether?*

- *What programs and other services are offered by the libraries for children and families? When are these programs and services available, and how long do they last? How many people can participate in these programs and services?*

STEP 2: MATHEMATIZE THE TOPIC.

Think about ways this topic involves mathematical ideas that you and your students have studied this year.

Example: Students' access to books and support for reading over the summer:

- *Number: counting books read, estimating number of books that can be read*

- *Measurement and computation: distance from home to the library, time needed to read a given number of books in comparison to the checkout period for books*

- *Fractional numbers: fraction of library programs for different ages of children, fraction of books available on a specific topic of interest in comparison to the total number of books*

STEP 3: PLAN A LEARNING ACTIVITY TO LAUNCH THE MATH INVESTIGATION.

Plan a learning activity to launch your investigation. You might, for instance, read a book related to the topic or share a math problem involving the issue you and your team have identified.

Example: Students' access to books and support for reading over the summer:

Share a flier from the public library about their summer programs. Ask students, "What do you notice? What do you wonder?"

The investigation itself will take place in three parts:

Here's what! Launch the investigation using the learning activity you and your team have planned to spark students' interest in the issue and activate schema. Explain to students that they will use mathematics to study this topic or issue and then they will decide how to take action in response to what they learn.

So what? As a class, brainstorm questions related to the issue that could be answered using mathematics. List these questions on the board. Decide together on a question or questions that the class will investigate, and design a plan for completing this investigation. Set aside time for a class discussion of what was learned.

Now what? Engage students in considering the implications of what they learned through their mathematical investigation and possible steps they can take in response to what they now understand. Decide on one or more actions that the class will commit to and then take action. At the end of the project, be sure to take time to reflect as a class on what students learned from the investigation about their mathematical powers and the role of mathematics in real life.

Examples: Students' access to books and support for reading over the summer:

Example 1—A class decides to investigate the proximity of the public library to students' homes. As they analyze data about distances from different neighborhoods to the library, available transportation options, and the library's hours of operation, they realize that it is not feasible for many families to use the library. They prepare a proposal to their principal to open the school library one evening a week.

Example 2—A class decides to investigate ways to help families build a habit of daily reading during the summer. They determine the number of books that might be read each week by an average family and the likely weight of these books. As a result of their learning, they solicit tote bags from a local grocery store for families to carry books to and from the library. Before distributing these tote bags, they prepare and include a flier outlining the benefits of summer reading, information about library hours and programs, and a reading log for families to use in building their family reading habits.

Let's Try It (10 minutes)

Implement the "Here's what! So what? Now what?" investigation that you and your colleagues planned. As students engage in this mathematics investigation, gather data about their use of the math habits. Choose one of the data-collection processes listed below.

OPTION 1: ANECDOTAL NOTES

Put a list of the math habits on a clipboard (Figure 11.2). As students engage in the investigation, take brief anecdotal notes about how students used the math habits, including the student's name and notes about what they did or said.

OPTION 2: STUDENT JOURNALING OR EXIT TICKETS

At the end of the investigation, pose questions such as these for students to respond to in writing:

- How did you use the math habits during this investigation?
- What did you learn about math from this investigation?
- What did you learn about yourself?

OPTION 3: CLASS DISCUSSION OR STUDENT INTERVIEWS

At the end of the investigation, hold a class discussion to give students a chance to reflect on their learning. You might audio record the class discussion to review again later. Pose questions such as follows:

- How did you use the math habits during this project?
- What did you learn about math from this project?
- What did you learn about yourself?

Alternatively, you might do one-to-one or small-group interviews of students using these same reflective questions.

Figure 11.2 • Habits of Mathematically Powerful People

I expect math to make sense.	I love challenging math problems.
I learn from math mistakes.	I see myself as a mathematician.
I talk about math.	I represent math in different ways.
I make math connections.	I look for and use patterns.

Images source: istock.com/Fourleaflover

ADDITIONAL PROFESSIONAL LEARNING EXPERIENCES

This month's additional professional learning experiences support educators in reflecting on the role of mathematics in their own lives and how they are helping their students to build a positive relationship with mathematics.

Math Talks: Real-World Math

This month's math talk is designed to support you and your team in thinking about the important role mathematics plays in real life.

Although this math talk focuses on a teacher issue, you can use these same "What do you notice? What do you wonder?" and the "Would you rather…?" number sense routines to help students think about the relevance of mathematics to their own lives. Be on the lookout for charts, graphs, and other real-world examples of mathematics related to your students' interests that you can use in these routines. You might also encourage students to bring in examples that they find.

Jenna Laib's Slow Reveal Graphs website (Laib, n.d., https://slowrevealgraphs.com/) offers graphs to help students of all ages think about math in the real world.

What do you notice? What do you wonder?

TEACHER SALARIES BY STATE			
STATE	2021–22 SALARY	2022–23 SALARY (ESTIMATED)	PERCENT CHANGE
Alabama	$55,834	$56,109	0.49%
Alaska	$74,167	$75,259	1.47%
Arizona	$56,775	$60,275	6.16%
Arkansas	$52,610	$53,317	1.34%
California	$88,508	$90,151	1.86%

TEACHER SALARIES BY STATE			
STATE	2021–22 SALARY	2022–23 SALARY (ESTIMATED)	PERCENT CHANGE
Colorado	$60,130	$61,907	2.96%
Connecticut	$81,185	$83,400	2.73%
Delaware	$65,647	$66,243	0.91%
District of Columbia	$82,523	$84,882	2.86%
Florida	$51,230	$52,362	2.21%

Source: Adapted from Will, 2023.

Would you rather . . .

Teach 187 days a year at your current salary?	Teach 197 days a year for a $4,000 increase in your annual salary?

 Available as a downloadable resource on the companion website.

Discuss the experience:

- NCTM (2020) stated that an important purpose of math education is "helping children learn to both understand and critique their world with mathematics" (p. 15). How could the use of math talks involving real-world examples of mathematics help to achieve these purposes?

- How could the use of math talks involving real-world examples of mathematics strengthen students' math identities and agency?

- How could you use the "What do you notice?" "What do you wonder?" and "Would you rather…?" number sense routines in your classroom?

Manipulatives and Models Matter: Math Power Tools

Across the school year, you and your colleagues have engaged in professional learning experiences involving a variety of mathematical tools. This month's Manipulatives and Models Matter activity offers the opportunity to reflect on these professional learning experiences along with your experiences using mathematical tools with your students. This reflection can support your math teaching community in designing a cohesive plan for math tool use next year.

Together with your team, reflect on the experiences you have had this year with mathematical tools.

- Which tools were new to you? How did these tools help you to think about mathematical ideas in different ways?

- What are the strengths of each of the tools you experienced? What kinds of purposes are they best suited for?

- What is your go-to math tool? What is one tool you would like to get to know better? Why?

- If your own math learning experiences in elementary school had involved more math tools, how might your current relationship with math be different?

According to Hiebert et al. (1997), mathematical tools serve three purposes in learning:

- Math tools help learners keep a record of their mathematical thinking.

- Math tools help learners communicate about their mathematical thinking.

- Math tools support and extend learners' mathematical thinking.

Together with your team, choose a manipulative model that you are familiar with. Think about ways you have seen students use this manipulative as a math tool for each of these three purposes. How did these experiences support their math learning?

Karp et al. (2021) reminded us that concrete, semi-concrete, and abstract representations can be important math learning tools. They told us, "How you choose the representations that your team will include in units is one of the critical decisions you have to make. Choosing a weak representation (or not choosing one at all!) can greatly affect student learning" (p. 52). These authors advised schools to make decisions about the tools that will be used at each grade level and across the school.

Together with your team, list the mathematical tools you used with students this year. Consider other math tools you may want to incorporate into your curriculum next year. Decide on the tools you all agree to use next year, your Math Power Tool list.

At a faculty meeting, have grade levels share their Math Power Tool lists. Consider whether the sequence of tools used across grade levels contributes to students' understanding of foundational math concepts and how this sequence might be strengthened.

Game Time

In this month's Game Time, we offer you a chance to reflect on the reasons for using games in the math classroom, to think about how you might use games in this last month of the school year to help students review and reflect on their math learning, and we share a personal favorite bonus game.

Take a few minutes to think about and discuss the following questions with colleagues:

- How have you used math games across the year?

- How did these games help you and your students to strengthen your math habits?

- What are the benefits of using games in the math classroom?

- What role might math games play in helping students review and reflect on their learning during this last month of school?

Here are some ideas for using games during this last month of the school year:

- Provide meaningful practice of essential math skills as well as some novelty in these last busy weeks of school by holding a Math Game-a-Thon. Students can choose from and play games they've learned across the year. Invite another class to join you for this fun event.

- Put together a booklet of games for students to play at home over the summer. You might have students work in collaborative groups to write and illustrate directions for their favorite math games.

- Plan games to use at an end-of-year family math celebration (see Let's Do Some Math Across Our School! on p. 385).

(Continued)

(Continued)

PIG

Pig is a simple but fun game for practicing mental math that involves mathematical probability.

Materials:

2 players

Two dice

Paper

Pencil

Challenge: Be the first person to reach 100.

Game directions:

1. On your turn, roll both dice and add the dots shown on each. You may roll the dice as many times as you wish, continuing to add their sum to your score.

2. That's the simple part, but there's a twist. If you roll a 1 on either dice, your score for that turn is 0 and your turn is over. And if you roll 1 on both dice, your total score for the whole game becomes 0 and your turn is over. So, you can roll as many times as you wish, but you're hoping not to roll any 1s.

3. When you decide to end your turn or when you roll one or two 1s, write your score on your paper and pass the dice to your partner. Your partner follows these same rules for their turn.

4. Continue taking turns until one player reaches a score of 100.

Ideas for simplifying the game for younger students:

• Play with only one die. Roll as many times as you wish but, if you roll a 1, your score is 0 and your turn ends.

• Play for a winning score less than 100 that is appropriate to your students' current understandings.

Ideas for making the game more complex:

• If you roll the same number on each die, your score for that roll is double the sum of the die. For instance, if you roll 5 and 5, your score for that roll would be 10 doubled or 20.

Discuss the experience:

- How did you engage in the math habits as you played Pig?

- How might you use Pig to support your classroom learning goals?

Inspired by the Pig game in *Math Games with Bad Drawings* (Orlin, 2022, pp. 261–262)

 Available as a downloadable resource on the companion website.

IN YOUR CLASSROOM

Classroom Story

Ms. Trinh's class had been working on writing their own math story problems involving planning for a special family meal and determining how to divide the foods that would be served so that every family member received a fair share. Ms. Trinh had chosen this project to give students a chance to apply their knowledge of division and fractions in real-world contexts. She also hoped to promote pride in individual students' cultural wealth and an appreciation for the different cultures represented within the classroom community. Students had enthusiastically written story problems involving dumplings, moussaka, churros, naan, and more.

During today's class, students shared and solved their problems in small groups. Ms. Trinh wrapped up the math lesson with a class discussion of the ways in which division and fractions had been used within the various problems. Then, she asked the students to think about their use of the math habits.

Ms. Trinh: *Wow! A lot of great math thinking happened today! Let's end math time by offering each other some math compliments. Who would like to start? Albert?*

Albert: *Lydia, I liked the table you made showing how you could share a lasagna with different numbers of people.*

Ms. Trinh: *Why do you like it?*

Albert: *It shows lots of problems. See, if there are six people, each gets one sixth of the lasagna, but if twelve people show up for dinner, they only get one twelfth.*

Ms. Trinh; *Thank you, Albert. Soo-Jin, did you want to give a math compliment?*

Soo-Jin: *Jaycob, when I couldn't remember what to call the different parts of a fraction, you helped me find the words numerator and denominator on* *the math word wall. You are really good with math words, and you're very helpful.*

Ms. Trinh: *Thank you, Soo-Jin. We have time for one more math compliment. Roxie?*

Roxie: *I wrote a story problem about sharing five Mandarin oranges between six people. Branson figured it out by drawing a picture. He made a connection to another problem he had solved about sharing almond cookies.*

Ms. Trinh: *Thanks, Roxie and everyone. You're really noticing each other's math strengths. That makes us a powerful math community!*

Ms. Trinh was proud of the many ways her students had worked together as learning partners and mathematical colleagues during the lesson. She smiled, knowing that this experience had strengthened their mathematical self-awareness and their appreciation of each other's mathematical insights.

Mathematics task inspired by Family Story Problems in Bartell et al. (2023).

Anchor Lesson

Mathematical Points of Power
Day 1: I Am Proud of My Mathematical Strengths!

WHAT?

This lesson will build students' awareness of their mathematical points of power. It will also help them appreciate the idea that the classroom community gives us all access to a variety of math strengths and that we're all smarter and stronger when we tap into and celebrate each other's mathematical points of power.

YOU NEED:

Eight pieces of chart paper featuring one habit of mathematically powerful people on each chart along with a graphic organizer as shown on page 383.

- I expect math to make sense.
- I love challenging math problems.
- I learn from math mistakes.
- I see myself as a mathematician.
- I talk about math.
- I represent math in different ways.
- I make math connections.
- I look for and use patterns.

I expect math to make sense.	
Looks like	Sounds like
• I read the problem several times. • I draw a picture to make sense of the problem. • I keep thinking.	• I ask myself "What do I know that can help me?" • I ask myself "How is this problem like other problems I've solved?" • I ask myself "What can I try?"

How I can use my strength to help others:

• I can ask them questions to help them think.

• I can tell them to keep trying.

HOW?

Part 1—Identifying Our Strengths

1. Post the eight habits of mathematically powerful people posters at different places around the room.

2. Remind students of the habits of mathematically powerful people that they have studied across the school year. Tell students that these habits are strengths they have and continue to grow. They are habits that people can use when they tackle challenging math problems. Tell students that within a classroom community, we all have different strengths. Say that when we use our personal strengths to help others, we all benefit.

3. Ask students to choose one of the eight habits that is a personal strength. Ask them to stand by the poster for that strength and, together with others who have also chosen that strength, talk about and record examples of what that strength looks like and sounds like and ways they might use their strength to help others.

REFLECT

Students have been learning about the habits of mathematically powerful people across the year. We can help students internalize these habits by making time for students to talk about their mathematical strengths.

Use the following discussion questions to reflect with students:

• What does it mean to be a mathematically powerful person?

• Why is it important to recognize your mathematical strengths?

• What is one math habit that you want to work to strengthen?

Day 2: Giving Strengths Compliments

WHAT?

This lesson will give students practice in noticing and naming other students' math strengths.

YOU NEED:

A sentence frame on chart paper: I saw you (name a math strength the student used). This is important because _____.

HOW?

1. Introduce the sentence frame.

2. Using examples students have generated on the charts they created the day before, model how they might use this sentence frame to name another student's math strength and describe why that strength is important.

3. Tell students that you will intentionally use this frame yourself to help them notice their own and others' math strengths. Say that during the closure of each day's math lesson, you'll invite them to use this frame to talk about the math strengths they notice in each other.

REFLECT

When we recognize, celebrate, and tap into each of our students' mathematical strengths, we truly create a powerful math community.

Use the following discussion questions to reflect with students:

- How can we use our math strengths to help each other?

- How can we help others to see their mathematical strengths?

- How do our math strengths help us to be a powerful math community?

LET'S DO SOME MATH ACROSS OUR SCHOOL!

This final month of the school year is the perfect opportunity to help students reflect on and celebrate their personal growth as mathematicians and for the school to reflect on and celebrate its growth as a mathematics learning community. As our valued partners in learning, parents and families should be a part of this celebration. An end-of-year math celebration is a chance to thank parents for their partnership, deepen parents' understanding of the math habits, and help parents learn more about how they can support their children's math learning.

You and your colleagues might structure this celebration as an evening event for students and their families. Feature students' Mathematical Me journals and portfolios from across the year prominently in this celebration. Here are some ideas to choose from. These suggestions may also spark additional ideas for your end-of-year math celebration:

- Have students choose several pieces of work from their portfolios that they are especially proud of. Display these prized artifacts of learning on a bulletin board or outside the classroom to spotlight students' mathematical brilliance.

- Display students' math portfolios and journals on students' desks for families to browse and enjoy. Make sticky notes or note paper readily available along with suggested sentence frames for family members to write compliments to student mathematicians.

- Display math anchor charts that you and your students created together across the year to show parents the important learning that occurred in your classroom since the start of school.

Alternatively, your school could celebrate students' mathematical growth in student-led parent conferences where students proudly show their parents select journal entries and examples of problem-solving work from their portfolios demonstrating specific aspects of their learning. They can discuss progress toward achieving their math learning goals and their thoughts about the mathematics they look forward to learning next year. You might draw from these ideas in planning and preparing for student-led conferences:

- Before the conferences, have students review their math journal entries and portfolio work to reflect on their learning and choose specific learning artifacts to share with their parents.

(Continued)

(Continued)

- Provide students with sentence frames to use in talking with parents about their learning:

  ○ This journal entry shows that I _____. This is important because _____.

  ○ I am proud of this piece of problem-solving work because _____.

  ○ One way that I want to continue growing as a mathematician is _____. This will help me _____.

- Give students time to practice their conference presentation with a peer. Practice partners might give each other feedback by sharing three things they enjoyed about the presentation and one suggestion for improvement.

In their book *Partnering with Parents in Elementary School Math*, Hillary Kreisberg and Matthew Beyranevand (2021) suggested that as parents arrive for their conference, you give them a simple handout outlining each person's role during the conference (Table 11.1).

Table 11.1 • Three-Way Mathematics Conference Roles

STUDENT, PARENT, AND TEACHER ROLES WITH CONFERENCES		
STUDENT	**PARENT**	**TEACHER**
• Reflects on their mathematics performance and effort, focusing on strengths. • Shares work samples and why they chose them. • Tells about self-identified areas of improvement. Describes goals they will set for themselves.	• Praises the student's efforts and identifies strengths in the student's work. • Asks questions about the shared work samples or content to better understand the mathematics. • Describes goals they want to see from the student if not mentioned. • Identifies ways to help the student reach their goals from the home front.	• Praises the student's effort and identifies strengths in the student's work. • Updates the parents and child as to next content topics and expectations. • Describes goals they want to see from the student if not mentioned.

Source: Reprinted with permission from Kreisberg & Beyranevand, 2021, p. 143.

A wealth of additional ideas for celebrating math learning with families can be found in Kreisberg and Beyranevand (2021).

TEACHING MOVE: MATH SUPER POWERS

As a class, create a chart that lists the strengths students identified in the anchor lesson (Table 11.2). Encourage students to continue adding their names below different strengths on the list as they gain proficiency and confidence with different math habits. Also, encourage students to use each other's strengths to support their own math learning. You might role-play with students how they could ask someone with a strength in perseverance for help in sticking with a particularly challenging problem or how they might ask someone with a strength in mathematical representation for ideas about how to create a graph.

Table 11.2 • Math Super Powers

PERSEVERANCE	CONSTRUCTING MODELS	COMMUNICATION
If I need someone to encourage me to keep working and not give up.	If I need someone to help me to show or explain my thinking using manipulatives or drawings.	If I need someone to help me understand the directions for the task. Or If I need someone to talk over an idea or possible strategy with me.
Then, I can go to _____ Mario Elena Max Tiffany Diamond	Then, I can go to _____ Jordyn Paris Jay	Then, I can go to _____ Bryce Hannah Sarita Jenna

Source: Reprinted with permission from Kobett and Karp, 2020, p. 193.

REFLECT

Kobett and Karp asserted that:

> All children should leave school being able to articulate what they do well in the activity of learning mathematics. Can all students do that? Their strengths need to be visible to them through obvious recognition and attention; then and only

(Continued)

(Continued)

then can these carefully communicated strengths be translated into Points of Power. (2020, p. 177)

- Are all of your students able to name their math strengths? How might you use a journaling activity or a quick student interview to find out?

- How is your classroom community growing stronger as a result of your focus on students' mathematical strengths?

TEACHING MOVE: END-OF-SCHOOL LETTER

Write an inspirational letter to give to your students on the last day of school celebrating the mathematical strengths they have grown across the school year and reminding them that these strengths will help them to do important things in and out of school, now and in the future.

You might include this inspirational quote by mathematician Frances Su:

Do you have problems you want to solve? Oceans you want to navigate? Patterns in the starry spheres of your life that you wish to understand? Then you can be a math explorer, since you were born with the human capacities to inquire and to reason. Dream of the sun, the moon, the stars, and the world that you will discover. Imaginative, creative, and unexpected enchantment await. (2020, p. 31)

This short news story illustrates the impact letters like these can have on your students: https://bit.ly/3P9COik (NBC News, 2023)

REFLECT

- How might your focus on your students' mathematical strengths this year shape their futures?

- What mathematical strengths have you grown this year?

CLASSROOM ROUTINE: MYSTERY MATHEMATICIANS

Set aside time during this last month of school for parent volunteers to visit the class and talk about how they use math in their work and daily life (see Family Newsletter on p. 396). You might invite one parent to talk to the class each day or set aside a day when all interested parents could share their math stories.

If parents cannot come to school during the day, offer to video record their presentation using Zoom. If parents speak a language other than English, their child can translate the presentation for the class.

Ask parents to plan for a 10-minute presentation. Suggest that they bring specific and, if possible, visual examples of ways that they use math to help students think about the importance of mathematics and how math is involved in many careers and life activities.

Prior to the day of the presentation, ask the mystery mathematician parent to provide three facts about themselves so that students can guess who they might be. Together with your students, brainstorm questions they might ask the mystery mathematicians to learn more about their work.

Inspired by Kreisberg and Beyranevand (2021, pp. 163–166).

STATION ACTIVITY: MATH STRENGTHS COUNTDOWN POEM

This station activity is designed to have students reflect on their year through writing. Have students use this pattern to write Math Strengths Countdowns as a way of celebrating their unique mathematical interests and powers (Tables 11.3 and 11.4):

Table 11.3 • Math Strengths Countdown 3–5

5 ways you have grown as a mathematician this year
4 math tools you like to use
3 favorite math topics
2 words that describe your math strengths
1 thing you want to learn about math next year

Here is a simplified version of this structure for younger students.

Table 11.4 • Math Strengths Countdown K–2

3 math strengths you are proud of
2 math activities you like
1 favorite thing about math

 Available as a downloadable resource on the companion website.

LITERATURE CONNECTIONS

***Small World* by Mercurio** (2019) (Grades K –2)

Inspired by women scientists with the Indian Space Research Organization, *Small World* tells the story of a young girl who experiences the beauty and wonder of the world with mathematical eyes. She grows up to become an astronaut and contemplates the magnitude and magnificence of our universe.

Activity: What are some ways that Nanda is mathematically powerful? What are some ways that you are mathematically powerful? How will you continue to stretch and grow your mathematical powers?

***Hidden Figures: The True Story of Four Black Women and the Space Race* by Shetterly** (2018) (Grades 3–5)

By the author of the original version of *Hidden Figures*, this children's biography tells the stories of four Black female mathematicians whose previously unrecognized contributions to NASA's space program were the key to its successful moon missions. Dorothy Vaughan, Mary Jackson, Katherine Johnson, and Christine Darden experienced enormous discrimination because of their race and gender but persevered in their efforts to use their mathematical talents to support space exploration.

Activity: Margot Lee Shetterly, the author of this book, wants us to see Dorothy Vaughan, Mary Jackson, Katherine Johnson, and Christine Darden as mathematically powerful role models. She stated, "Dorothy, Mary, Katherine, and Christine knew one thing: with hard work, perseverance, and a love of math, *anything* was possible" (emphasis in original; quote appears on the 29th page). What do you admire about these four mathematicians? How can mathematics empower you to do important things both now and in the future?

SPOTLIGHT ON BRAIN SCIENCE

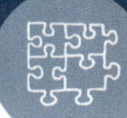

Our students' lives are different from our own experiences growing up. Students are expected to learn more and to learn at a faster pace than ever before. They have more choices and must make more decisions than in the past. Their futures will undoubtedly involve even more complexity and information management. Khan (2012) of Khan Academy stated, "Since we can't predict exactly what today's young people will need to know in ten or twenty years, *what* we teach them is less important than *how* they learn to teach themselves" (p. 180).

The executive functions of our brains control our higher level brain activities, including our ability to set and reach goals. This month's spotlight on brain science helps students to know that, even though their executive functions will continue to develop into adulthood, they can strengthen these higher level brain functions through practice. It suggests a simple action plan students can use to set goals for themselves as growing mathematicians and create plans to help achieve these goals.

You can share this information with students in a mini lesson or a station activity. You might also send this reproducible home along with the Family Newsletter to help families talk about these important ideas.

WHEN WE SET GOALS, WE STRENGTHEN OUR BRAIN FUNCTIONS.

The executive functions of your brain help you control your thoughts, feelings, and actions. These functions will continue to develop until you are an adult, but you can help these brain functions to grow stronger now through practice. You use your executive functions when you set learning goals for yourself as a mathematician.

Image source: istock.com/Elena Chiplak

How do you want to continue growing as a mathematician?

I.   Think about some of the ways you've grown as a mathematician this year. What are you proud of? What mathematics are you looking forward to learning about next year? How can you stretch your math powers? Choose one math learning goal that is important to you. Write it down and keep your goal somewhere where you will look at it often.

2.   What can you do to begin working toward this goal? What might you do this summer to start on this learning? Who might be able to help you?

3.   Imagine yourself in the future when you have achieved your goal. What will you do and say to show that you are proud of your learning?

As a powerful mathematician, you can think about and decide what math you want to learn next and then take steps to reach your goal.

Choose one of the following ways to think more about this important information:

I.   Write down why these ideas are important to you and how you want to put them to work in your life.

2.   Talk to a friend about these ideas. Decide together how you will put them to work.

3.   Share these ideas with someone at home. Tell them why these ideas are important to you and how you plan to use them.

 Available as a downloadable resource on the companion website.

SPOTLIGHT ON EQUITY: CELEBRATE CULTURAL WEALTH

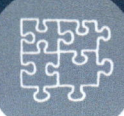

If we want all of our students to recognize themselves as mathematicians and to see math as a tool for accomplishing things that they care about, we need to provide opportunities for students to make connections between their lived experiences and mathematics.

Particularly when our own cultural backgrounds and life experiences are different from those of our students, our preparation for teaching must include learning about our students' funds of knowledge and the rich wealth of experiences, perspectives, and strengths they have through their families and communities.

TRY IT

With colleagues or on your own, drive through the neighborhoods where your students live. As you drive, make note of aspects of your students' cultural wealth that you might incorporate in math tasks during this last month of school or next school year. Notice parks, stores, places of worship, means of transportation, and so on. Jot down names of streets, buildings, and other attractions.

Afterward, write up several math tasks using the details you have gathered. As you use these tasks in class, notice students' reactions. Talk to students about the idea of cultural wealth, and tell them you are working on designing math tasks that incorporate examples of their cultural wealth. Ask for feedback from students about ways to make tasks more authentic. Invite students to write their own math tasks that illustrate how math is a part of their life outside of school.

Inspired by the Eyes! Camera! Action! Adventure activity in *Strengths-Based Teaching and Learning in Mathematics* (Kobett & Karp, 2020, pp. 135–136)

MATHEMATICAL ME: STUDENT JOURNAL AND PORTFOLIO

Student Journals:

Have students respond to some or all of the following questions in their journals:

- What are my math strengths?

- How can I use math to do important things in and out of school?

- What are we celebrating about our powerful math community?

Older students can write their responses in the Mathematical Me journals. Younger students can draw a picture showing their response to a question, or you might gather responses during a class discussion or quick one-on-one interviews.

Review students' responses to monitor their mathematical identity and agency. You might summarize this data by tallying or graphing the different ideas students offer.

Be sure to share and discuss this data with your class. Ask, "What does this data tell us? and "What are some things we can do to continue growing as powerful mathematicians?"

Student Portfolios:

Have students choose a piece of work that shows one or more of their math strengths. On a sticky note, have them write the date and complete this sentence frame:

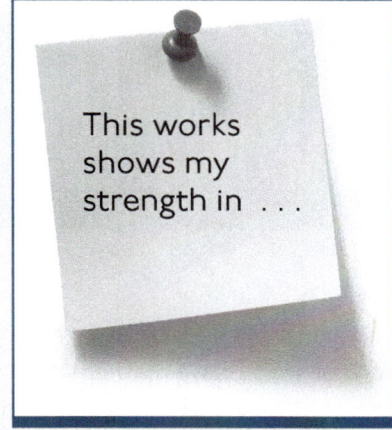

Source: Istock.com/Thammask-Chuenchom

REVISITING THE ATTITUDE SURVEY

Readminister the Attitude Survey that students took in August and again in December (see pp. 56–59). Guide students in reflecting on and comparing their results across the year. On a sticky note, have them complete the sentence frame:

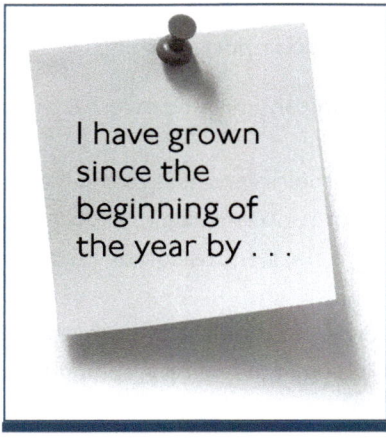

I have grown since the beginning of the year by . . .

Source: Istock.com/Thammask-Chuenchom

Be sure to include all three surveys and sticky notes in students' portfolios.

FAMILY NEWSLETTER: WE ARE POWERFUL MATHEMATICIANS!

WHAT WE'RE LEARNING

This month we're learning that we all have mathematical strengths, things we do well that we can draw on when we engage in challenging mathematics. We can use our math strengths to learn more math. We can also develop additional mathematical strengths.

Here are the mathematical strengths that your child has been growing all year:

- I expect math to make sense.
- I love challenging math problems.
- I learn from math mistakes.
- I see myself as a mathematician.

- I talk about math.
- I represent math in different ways.
- I make math connections.
- I look for and use patterns.

WHY IT'S IMPORTANT

When your child recognizes their mathematical strengths and knows they can use these strengths to tackle challenging problems and continue building their mathematical understandings and skills, they will be successful in school mathematics and can use math as a tool in real life.

HOW YOU CAN HELP

When you notice your child demonstrating one of these math strengths, take a minute to compliment your child and remind them why that strength is valuable. In school, we're using this sentence frame for these strengths compliments:

I saw you (name a math strength that your child used). That strength is important because _____.

Another way you can help is by talking about the ways you use mathematics at home and at work, your personal math strengths. We would love to have you share this personal math story with your child's class during our Mystery Mathematician time. Please contact your child's teacher if you are willing to do a brief presentation to your child's class about how you use math in real life.

 Available as a downloadable resource on the companion website.

CHECKING IN ON OUR LEARNING: WE ARE POWERFUL MATHEMATICIANS!

- What are my mathematical strengths?
- How can I use mathematics to do important things in and out of school?
- What are we celebrating about our powerful math community?

This month's learning focus is about taking pride in the fact that we are powerful mathematicians and knowing that our math habits help us accomplish things we care about; this idea is central to students' and teachers' mathematical identities and agency. As teachers and students celebrate their math learning and set new math learning goals, they create momentum for self-directed math learning and continuous improvement of the school's mathematics program.

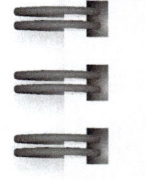

MATHEMATICAL ME: EDUCATOR JOURNAL

- How has this month's learning focus supported your students' mathematical growth?

- How has this month's learning focus supported your growth as a math teacher?

- How has this month's learning focus supported your school's growth as a powerful math community?

LOOKING AHEAD

The downloadable resources available at our companion website (**https://qrs.ly/vqfn1s2**) include a bonus chapter resource called "Chapter 12—This Summer's Focus: Powering Up for a New School Year!" to support you and your teaching community in reflecting on the educator learning that has occurred over the past year and articulating your school community's vision for its mathematics instructional program. This resource will also help you to consider options for next year's practice-based professional learning to support your school's continued improvement of mathematics teaching and learning.

Questions for Reflection

- What are you most proud of in your own growth as a mathematics teacher this year? Why?

- What school-wide improvements in math teaching and learning are you most proud of? Why?

- What parts of your math teaching practice are you eager to continue refining next year? What support will you need for this professional learning work?

- What improvements in the school's mathematics instructional program would you like to see the school community tackle together next year? Why?

Epilogue

"I touch the future. I teach."

—credited to Christa McAuliffe,
Teacher and Astronaut

Infinity is the idea of endlessness. It is a concept that goes beyond what we can see with our eyes; it is challenging to imagine. Counting can help us to think about the idea of infinity. Imagine counting a really large quantity of objects, like stars in the universe. No matter how enormous a number we count, we can always count one more, and one more, and one more endlessly. Even if we count forever, we will never reach infinity. The idea of infinity is useful in solving important real-world problems, including problems related to astronomy and space travel. Mathematicians regularly put the idea of infinity to work in thinking through problems involving complex predictions (Campbell, 2022).

Your efforts this year have set in motion powerful dynamics that are already impacting your students' readiness and eagerness to learn more math. Your teaching has also shaped your students' beliefs about the role that mathematics can play in their lives. The understandings, abilities, mindsets, and aspirations you helped your students to grow this year will, across time, touch and shape numerous other peoples' beliefs about the nature of mathematics and their relationship with math. Like the idea of infinity, we can only imagine the never-ending impact of the important work you do each day as a math educator. You are our hero!

—Holly and Sue

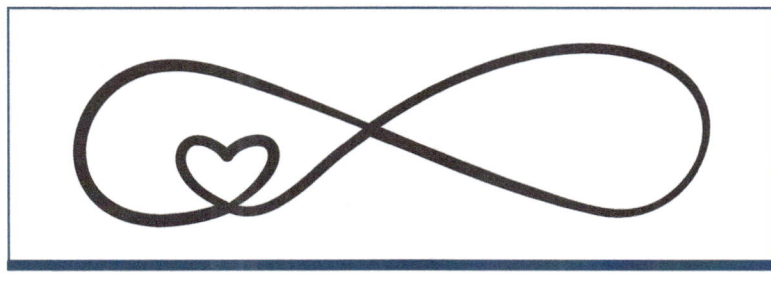

Image source: Istock.com/Daniil Chaban

We would be honored to connect with you. Please consider sharing your school's math learning story on social media (#PowerUpMath @Holly_Burwell @SueChapmanLearn) and with us directly at hollyburwell @inspiredmathematicsmt.com and SueChapmanLearning@gmail.com.

Appendix
Additional Professional Resources

In a powerful math community, educators know that teaching is a learning profession and that efforts to strengthen their instructional craft never end. How will you build on the growth you and your community have achieved this year?

Here is a short list of outstanding professional books to support your school's next steps in powering up mathematics teaching and learning.

- ***Cultivating Mathematical Hearts: Culturally Responsive Mathematics Teaching in Elementary Classrooms*** by Maria del Roario Zavala and Juia Mari Aguirre (2024, Corwin)

  Designed to help teachers place students at the center of their math instruction, this book introduces readers to big ideas about culturally responsive mathematics teaching and helps them apply these ideas to a variety of classroom contexts. Packed with classroom stories, professional learning activities, and reflective exercises, *Cultivating Mathematical Hearts* will help your school to humanize math education and power up students' and teachers' visions of who we are as mathematical beings.

- ***The Imperfect and Unfinished Math Teacher: A Journey to Reclaim Our Professional Growth*** by Chase Orton (2022, Corwin)

  This book is a joyous and inspiring exploration of what it means to be a math teacher. It offers educators a road map to improving their instructional practice in community with fellow teachers. Thought-provoking discussion prompts and reflective exercises support readers in thinking deeply about mathematics teaching and learning. Orton's teaching stories and his invitation to write our own math teacher stories as we support students in authoring their math learning stories will make this a favorite mentor text for your teachers.

- ***The Math Pact: Achieving Instructional Coherence Within and Across Grades*** by Karen Karp, Barbara Dougherty, and Sarah Bush (2021, Corwin)

  The "math pact" is a mathematics whole-school agreement developed collaboratively by a school community to ensure that all students receive cohesive and highly effective math instruction on a daily basis. This guidebook takes educator teams through the process of establishing consistent and aligned expectations for instructional practices related to vocabulary, mathematical notation, representations, rules, and generalizations within and across classrooms and grade levels. This book is another great choice for powering up your math community.

- ***Mathematizing Your School: Creating a Culture for Math Success*** by Nicki Newton and Janet Nuzzie (2019, Routledge)

  This book is a collaboration between math consultant Dr. Nicki Newton and district math leader Janet Nuzzie. It offers practical suggestions and tools to strengthen your mathematics program at the classroom, school, and district levels.

- ***Partnering with Parents in Elementary School Math: A Guide for Teachers and Leaders*** by Hilary Kreisberg and Matthew Beyranevand (2021, Corwin)

  This book is a practical but comprehensive guide to building a strong math learning alliance with families. From understanding what parents need from schools, to establishing effective communication systems and channels, to planning and implementing family events to support math learning, this book provides a school community with an array of understandings and tools to partner with parents in growing students as mathematicians.

- ***Productive Math Struggle: A 6-Point Action Plan for Fostering Perseverance*** by John SanGiovanni, Susie Katt, and Kevin Dykema (2020, Corwin)

  This "action plan for fostering perseverance" directly tackles the counterintuitive notion that struggle is essential to real learning. The authors brilliantly show educators how to help students to develop the habits and dispositions necessary to relish and benefit from productive math struggle. Organized into six concrete action steps, this book would be a valuable yearlong book study for a school community.

- ***Strengths-Based Teaching and Learning in Mathematics: 5 Teaching Turnarounds for Grades K-6*** by Beth Kobett and Karen Karp (2020, Corwin)

  Another great choice for a yearlong school-wide book study, *Strengths-Based Teaching and Learning in Mathematics* helps educators shift from a focus on students' math deficits toward the understanding that all people are mathematicians by nature and that teachers are most effective when they identify and build on their students' mathematical strengths and proficiencies. Filled with practical and easy-to-implement ideas appropriate for all elementary grades, this book can help you to continue powering up your math community.

Glossary

Abstract representation: a mathematical representation using numbers, symbols, equations, and words.

Advancing questions: teacher questions designed to deepen student thinking and learning.

Assessing questions: teacher questions designed to surface student thinking.

Clarifying moves: verbal moves teachers can use to support students in clarifying or refining their thinking.

Classroom investigation: a teacher-driven approach to professional learning in which a teacher or teacher team identifies a student learning need, articulates a student learning goal, designs a plan for achieving that goal, implements the plan, and then measures its impact on student learning.

Cognitive dissonance: the recognition that something doesn't make sense.

Concrete representation: a mathematical representation created with manipulative models.

Concrete, representational, abstract (CRA) approaches to learning: a process of learning that uses concrete and representational mathematical models to support the learning of abstract concepts.

Cultural wealth: strengths people acquire as a result of their families' and communities' cultures that make them unique as learners and are valuable assets to the class community.

Dependent learners: learners who do not expect math to make sense and do not see themselves as capable of making sense of mathematics. Dependent learners lack agency. They do not yet have strategies for tackling and persevering with challenging mathematics.

Eliciting moves: verbal moves teachers can use to support students in articulating their thinking.

Equitable participation: an exchange of mathematical ideas in which all students have the opportunity and support needed to share their thinking and ask questions.

Fixed mindset: the belief that learning is the result of innate ability rather than of effort.

Focusing questions: teacher questions designed to spark critical thinking and help students to notice and make sense of important mathematical ideas (e.g., "What do you know about 0.3?").

Fractal: a never-ending pattern that grows incrementally by repeating a mathematical process again and again.

Funds of knowledge: knowledge people acquire as a result of their life experiences that make them unique as learners and that are valuable assets to the class community.

Funneling questions: teacher questions designed to elicit an answer to a question that the teacher considers correct (e.g., "Which digit is in the tenths place?").

Growth mindset: the belief that learning is the result of effort rather than of innate ability.

Habit: an automatic behavior, often engaged in without conscious thought.

Habits of mathematically powerful people: a set of practices used in mathematics that grow a person's agency and identity.

Independent learners: self-directed learners who take initiative in their own learning. Independent learners have agency. They have strategies for tackling and persevering with challenging mathematics.

Infinity: the concept of being endless or infinite. Counting and numbers are sometimes used to illustrate the idea of infinity. We can count forever without reaching infinity.

KASAB: an acronym for five essential professional learning outcomes: knowledge, attitudes, skills, aspirations, and behaviors.

Lesson study: a teacher-driven approach to professional learning in which teachers collaborate in planning, observing, reflecting, and refining a lesson focused on an identified student learning goal.

Looking for patterns: the process of looking across several problems, objects, or numbers to look for similarities.

Making connections: the process of seeking out how two or more things relate to each other.

Manipulative: concrete model of a math concept that students can manipulate physically.

Mantra: a statement that is repeated frequently to help internalize big ideas.

Math agency: a person's self-directedness in learning mathematics and using mathematics in real life.

Math café: a mini professional learning experience.

Math content: the mathematical concepts and skills that are taught in each grade.

Math discourse: the exchange of mathematical ideas in a classroom.

Math identity: a person's beliefs about their capacity for understanding mathematical ideas and the importance of mathematics in their life.

Math practices: the habits of mathematical thinking or processes for engaging in math that are grown across K–12.

Math talk: a brief class discussion of a math problem or idea that allows students to practice mathematical reasoning and communication.

Math teaching community: a team of educators who work together to maximize the math learning of all students across the school community.

Mathematical points of power: mathematical strengths that students can draw on in tackling challenging problems.

Metacognition: thinking about our own thinking, self-talk.

Mistake: a strategy or solution that produces an incorrect result; a natural and needed part of learning.

Model: a representation of a mathematical idea. Typically, the term refers to a visual representation of a mathematical idea, but it can also refer to an individual's mental model.

Number sense: the ability to reason flexibly about numbers and number relationships.

Number talk: a 5–15-minute number sense routine in which students have discussion and solve computation problems mentally.

Operation sense: an understanding of the meaning and uses of the four number operations: addition, subtraction, multiplication, and division.

Orienting moves: verbal moves teachers can use to support students in thinking about and responding to others' ideas.

Perseverance: the ability to push through confusion and struggle in learning into understanding. Perseverance can be thought of as self-scaffolding. The ability to persevere can be strengthened through practice and coaching.

Practice-based professional learning: educator learning that takes place in or close to the classroom, allowing teachers to try out new instructional practices and examine the impact of these practices on student learning.

Probing moves: verbal moves teachers can use to follow up on a student's initial response in order to deepen thinking.

Problem posing: posing questions or problems related to real-life situations that can be answered or solved mathematically.

Productive struggle: cognitive effort to make sense of and solve challenging mathematics problems. Productive struggle is essential to learning.

Professional learning cycle: an iterative professional learning process in which teachers engage in reflection, learning, planning, implementing their new learning, and finally measuring the impact of what they've tried on student learning.

Representational fluency: the ability to flexibly use and translate between different mathematical representations.

Respectful discourse: an exchange of mathematical ideas in which each student's voice is valued.

Self-fulfilling prophesy: the belief that what you think and talk about directly affects your actions.

Semi-concrete representation: a two-dimensional representation of a mathematical idea. Semi-concrete representations commonly used in elementary school include five-frames, ten-frames, hundreds charts, traditional and open number lines, arrays and area models, multiplication charts, diagrams, tables, and graphs. Also called a pictorial representation.

Shame: the painful feeling that is accompanied by the awareness that a person has made a mistake.

Standards for Mathematical Practices (SMPs): a set of eight practices also known as "habits of mathematical thinking" derived from the Common Core State Standards that were released in 2010.

Strengths-based language: language describing people by their abilities as opposed to by their challenges.

Talk facilitation moves: verbal moves teachers can use to support students' mathematical reasoning and communication.

Talking-to-learn moves: verbal moves students can use to share their math thinking, actively listen to others, and co-construct mathematical knowledge as a member of a math learning community. Talking-to-learn moves include explaining and justifying, connecting, conjecturing, critiquing, and questioning.

Tool: a mathematical model becomes a tool when its structure is understood, allowing it to be used for mathematical thinking and problem-solving.

Unproductive struggle: cognitive struggle in which a learner does not have the internal resources or external supports to push through struggle into understanding and therefore gives up.

References

Aguilar, E. (2020). *Coaching for equity: Conversations that change practice.* Jossey-Bass.

Aguilar, E., & Cohen, L. (2022). *The PD book: 7 habits that transform professional development.* Jossey-Bass.

Aguirre, J., Mayfield-Ingram, K., & Martin, D. B. (2013). *The impact of identity in K-8 mathematics: Rethinking equity-based practices.* National Council of Teachers of Mathematics.

Alexie, S. (2016). *Thunder boy Jr.* Little, Brown Books for Young Readers.

Ani, K. (2021). *Dear citizen math: How math class can inspire a more rational and respectful society.* Damascus Rodeo.

Barrera, F., & Santos, M. (2001). Students' use and understanding of different mathematical representation of tasks in problem-solving instruction. In R. Speiser, C. A. Maher, & C. N. Walter (Eds.), *Proceeedings of the twenty-third annual meeting of the North American chapter of the International Group for the Psychology of Mathematics Education* (Vol. 1, pp. 459–466).

Bartell, T. G., Yeh, C., Felton-Koestler, M. D., Berry, R. Q., & Lawler, B. (2023). *Upper elementary mathematics lessons to explore, understand, and respond to social injustice.* Corwin.

Bernstein, M. F. (2019, November 13). Mind of a mathematician. *Princeton Alumni Weekly.* https://paw.princeton.edu/article/mind-mathematician

Boaler, J. (2015). *Mathematical mindsets: Unleashing students' potential through creative math, inspiring messages and innovative teaching.* Jossey-Bass.

Boaler, J. (2019). *Limitless mind: Learn, lead, and live without barriers.* HarperOne.

Boaler, J. (2022). *Mathematical mindsets: Unleashing students' potential through creative mathematics, inspiring messages and innovative teaching.* Jossey-Bass.

Bor, D. (2012). *The ravenous brain: How the new science of consciousness explains our insatiable search for meaning.* Basic Books.

Bourassa, M. (2013). *Which one doesn't belong?* WODB. https://wodb.ca/

Bresser, R., & Holtzman, C. (2018). *Math workshop essentials: Developing number sense routines, focus lessons, and learning stations.* Math Solutions.

Brown, B. (2013, January 15). *Shame vs. guilt.* https://brenebrown.com/articles/2013/01/15/shame-v-guilt/

Bruner, J. (1960). *The process of education.* The President and Fellows of Harvard College.

Burns, M. (1982). *Math for smarty pants.* Scholastic.

Burns, M. (2007). *About teaching mathematics: A K–8 resource* (3rd ed.). Math Solutions.

Burns, M. (2015a). *About teaching mathematics: A K–8 resource* (4th ed.). Math Solutions.

Burns, M. (2015b, April 28). *An unusual word problem: "Dealing in horses".* Marilyn Burns Math. https://marilynburnsmath.com/general-interest/the-dealing-in-horses-problem/

Burns, M. (2015c, October 1). *Can you KenKen?* Marilyn Burns Math. https://marilynburnsmath.com/games/can-you-kenken/

Burns, M. (2020). *Welcome to math class: A collection of Marilyn's favorite lessons, grades K-6.* Math Solutions.

Burns, M. (2022). *About teaching mathematics: A K–8 resource* (1st ed.). Heinemann.

Burns, M. (2023). *Conversation 9: Why play games in math class?* [Audio podcast]. Marilyn Burns Math. https://marilynburnsmath.com/math-podcasts/conversation-9-why-play-games-in-math-class/

Campbell, S. C. (2022). *Infinity: Figuring out forever.* Astra Young Readers.

Carlos, M. (2020, October 7). *Global patterns.* https://www.michellecarlos.com/post/global-patterns

Chapin, S. H., O'Connor, C., & Anderson, N. C. (2022). *Talk moves: A teacher's guide for using classroom discussions in math, grades K-6.* Heinemann.

Chapman, S., & Mitchell, M. (2021). *MathVentures: 33 teacher-coach investigations to grow students as mathematicians.* Math Solutions.

Clear, J. (2018). *Atomic habits: An easy & proven way to build good habits & break bad ones.* Penguin.

Cole, S. (2012, December 8). *Michael Jordan "failure" commercial HD 1080p* [Video]. YouTube. https://www.youtube.com/watch?v=JA7G7AV-LT8

Confer, C. (2017). *Sizing up measurement: Activities for grades 3-5 classrooms.* Math Solutions.

Coulson, A. (2021). *Look, grandma! Ni, Elisi!* Charlesbridge.

Crespo, A. (2020). *Lia & Luís: Who has more?* Charlesbridge.

Crespo, A. (2023). *Lia & Luís: Puzzled!* Charlesbridge.

Crespo, S., Celedón-Pattichis, S., & Civil, M. (2018). *Access & equity: Promoting high-quality mathematics, grades 3-5.* National Council of Teachers of Mathematics.

De Francisco, C., & Burns, M. (2002). *Teaching arithmetic: Lessons for decimals and percents.* Math Solutions.

Fennell, F. (2006, September). *Representation—Show me the math!* National Council of Teachers of Mathematics. https://www.nctm.org/News-and-Calendar/Messages-from-the-President/Archive/Skip-Fennell/Representation%E2%80%94Show-Me-the-Math!/

Fredrickson, B. L. (2009). *Positivity: Discover the upward spiral that will change your life.* Harmony Books.

Gojak, L. M. (2013, October 3). *Making mathematical connections.* National Council of Teachers of Mathematics. https://www.nctm.org/News-and-Calendar/Messages-from-the-President/Archive/Linda-M_-Gojak/Making-Mathematical-Connections/

Gonzalez, L. (2023). *Bad at math? Dismantling harmful beliefs that hinder equitable mathematics education.* Corwin.

Hammond, Z. (2014). *Culturally responsive teaching and the brain: Promoting authentic engagement and rigor among culturally and linguistically diverse students.* Corwin.

Hattie, J., Fisher, D., & Frey, N. (2017). *Visible learning for mathematics: What works best to optimize student learning.* Corwin.

Hesselbert, J. (2020). *Pitter pattern.* Greenwillow Books.

Hiebert, J., Carpenter, T. P., Fennema, E., Fuson, K. C., Wearne, D., Murray, H., Olivier, A., & Human, P. (1997). *Making sense: Teaching and learning mathematics with understanding.* Heinemann.

Hilton, H. (2019). *Look: I'm a mathematician.* DK Publishing.

Hobai, I. (2020). *A whale of a mistake.* Page Street Kids.

Hopkins, L. B. (Ed.). (2001). *Marvelous math: A book of poems.* Simon & Schuster Books for Young Readers.

Hua, H. (2023, January 15). *"How many?" images.* https://howiehua.wordpress.com/2023/01/15/how-many-images/

Huinker, D., & Bill, V. (2017). *Taking action: Implementing effective mathematics teaching practices, K-5.* National Council of Teachers of Mathematics.

Joboaler.org. (n.d.). *Struggly. Creating limitless learners.* https://joboaler.org/struggly/

Johnston, P. H. (2012). *Opening minds: Using language to change lives.* Stenhouse.

Kane, D. (2020, October 6). How to help students see themselves as mathematicians. *Edutopia.* https://www.edutopia.org/article/how-help-students-see-themselves-mathematicians/

Kaplinsky, R. (n.d.). *Open middle.* Open Middle. https://www.openmiddle.com/

Karp, K. S., Dougherty, B. J., & Bush, S. B. (2021). *The math pact: Achieving instructional coherence within and across grades.* Corwin (a joint publication with National Council of Teachers of Mathematics).

Khan, S. (2012). *The oneworld schoolhouse: Education reimagined.* Hachette UK.

Killion, J., Sommers, W. A., & Delehant, A. (2023). *Elevate school-based professional learning.* Solution Tree.

Knight, A. J. (2021). *Usha and the big dipper.* Charlesbridge.

Kobett, B., & Karp, K. (2020). *Strengths-based teaching and learning in mathematics: Teaching turnarounds for grades K–6.* Corwin.

Kohn, A. (2013, October 29). *A dozen essential guidelines for educators.* https://www.alfiekohn.org/blogs/dozen-essential-guidelines-educators/

Kreisberg, H., & Beyranevand, M. (2021). *Partnering with parents in elementary school math: A guide for teachers and leaders.* Corwin.

Laib, J. (n.d.). *Slow reveal graphs.* https://slowrevealgraphs.com/

Lambert, R. (2024). *Rethinking disability and mathematics: A UDL math classroom guide for grades K–8.* Corwin.

Lawrence, B., Sudeikis, J., Hunt, B., Kelly, J., Ingold, J., & Wrubel, B. (Executive Producers).

(2020–present). *Ted Lasso* [TV series]. Ruby's Tuna Inc., Doozer, Universal Television, Warner Bros. Television Studios.

Leedy, L. (2008). *Missing math: A number mystery.* Two Lions.

Leedy, L. (2012). *Seeing symmetry.* Holiday House.

Liljedahl, P. (2021). *Building thinking classrooms in mathematics, grades K–12: 14 teaching practices for enhancing learning.* Corwin.

MacDonald, E. B. (2023). *Intentional moves: How skillful team leaders impact learning.* Corwin.

Markowitz, N. L., & Bouffard, S. M. (2020). *Teaching with a social, emotional, and cultural lens: A framework for educators and teacher educators.* Harvard Education Press.

Mashup Math. (2023). *Super silly math jokes for kids.* Mashup Math LLC.

Matthews, L. E., Jones, S. M., & Parker, Y. A. (2022). *Engaging in culturally relevant math tasks: Fostering hope in the elementary classroom.* Corwin.

McConchie, L., & Jensen, E. (2020). Teaching to the whole brain. *ASCD, 77*(8). https://www.ascd.org/el/articles/teaching-to-the-whole-brain

McHugh, M. L. (2023). *Bringing project-based learning to life in mathematics, K–12.* Corwin.

Mercurio, I. (2019). *Small world.* Abrams Books.

Meyer, D. (2023, June 1). AI does not fit the shape of schooling. *Mathworlds.* https://danmeyer.substack.com/p/ai-does-not-fit-the-shape-of-schooling?utm_source=post-email-title&publication_id=239666&post_id=125203802&isFreemail=true

Mirror Neurons. (n.d.). PBS Learning for Teachers. https://az.pbslearningmedia.org/resource/hew06.sci.life.reg.mirrorneurons/mirror-neurons/

National Council of Teachers of Mathematics (NCTM). (2000). *Principals and standards for school mathematics.*

National Council of Teachers of Mathematics (NCTM). (2014). *Principals to actions: Ensuring mathematical success for all.*

National Council of Teachers of Mathematics (NCTM). (2020). *Catalyzing change in early childhood and elementary mathematics: Initiating critical conversations.*

National Governors Association Center for Best Practices & Council of Chief State School Officers. (2010). *Mathematics standards.* Common Core State Standards Initiative. http://www.corestandards.org/Math

National Research Council & Mathematics Learning Study Committee. (2001). *Adding it up: Helping children learn mathematics.* National Academies Press.

NBC News. (2023, March 5). *Michigan school staff uplift students through letters.* YouTube. https://www.youtube.com/watch?v=EYbBlWXSItA

Nottingham, J. (2017). *The learning challenge: How to guide your students through the learning pit to achieve deeper understanding.* Corwin.

Oakley, B., Rogowsky, B., & Sejnowski, T. J. (2021). *Uncommon sense teaching: Practical insights in brain science to help students learn.* TarcherPerigee.

Orlin, B. (2022). *Math games with bad drawings: 75¼ simple, challenging, go-anywhere games— and why they matter.* Black Do & Leventhal.

Orton, C. (2022). *The imperfect and unfinished math teacher: A journey to reclaim our professional growth.* Corwin.

Parr, T. (2014). *It's okay to make mistakes.* Little Brown Books.

Parrish, S. (2022). *Number talks: Helping children build mental math and computation strategies.* Heinemann.

Parrish, S., & Dominick, A. (2016). *Number talks: Fractions, decimals, and percentages.* Math Solutions.

Petersen, J. (2022). *Math games for number and operations and algebraic thinking: Games to support independent practice in math workshops and more.* Heinemann.

Picha, G. (2022). *Conferring in the math classroom: A practical guidebook to using 5-minute conferences to grow confident mathematicians, K-5.* Stenhouse.

Ray, M. (2013). *Powerful problem solving: Activities for sense making with the mathematical practices.* Heinemann.

Reid, M. (2021). *Maryam's magic: The story of mathematician Maryam Mirzakhani.* Harper Collins.

Reinhart, S. C. (2000). Never say anything a kid can say! *Mathematics Teaching in the Middle School, 5*(8), 478–473.

Rhodes, S., Mickle Moldavan, A., Smithey, M., & DePiro, A. (2023). Five keys for growing confident mathematics learners. *Mathematics Teacher: Learning and Teaching PK-12, 116*(1), 8–15. https://doi.org/10.5951/MTLT.2022.0225

Safir, S., & Dugan, J. (2021). *Street data: The next-generation model for equity, pedagogy, and school transformation.* Corwin.

SanGiovanni, J. J., Katt, S., & Dykema, K. J. (2020). *Productive math struggle: A 6-point action plan for fostering perseverance.* Corwin.

SanGiovanni, J. J., Katt, S., Knighten, L. D., & Rivera, G. (2022). *Answers to your biggest questions about teaching elementary math.* Corwin.

Seeley, C. (2015). *Faster isn't smarter: Messages about math teaching and learning in the 21st century.* Math Solutions.

Shaskan, T. S. (2009). *If you were a plus sign.* Picture Window Books.

Shetterly, M. L. (2018). *Hidden figures: The true story of four Black women and the space race.* Harper.

Shirley, D., & Hargreaves, A. (2024). *The age of identity: Who do our kids think they are… and how do we help them belong?* Corwin.

Short, J. B., & Hirsh, S. (2023). *Transforming teaching through curriculum-based professional learning: The elements.* Corwin.

Singh, S., & Brownell, C. (2019). *Math recess: Playful learning in an age of disruption.* IMPress.

Smith, D. (2021). *100 inspirational quotes by Maya Angelou.* Self-Published.

Smith, D. J. (2020). *If the world were a village: A book about the world's people.* Kids Can Press.

Smith, M. S., & Stein, M. K. (2018). *5 practices for orchestrating productive mathematics discussions* (2nd ed.). National Council of Teachers of Mathematics.

Su, F. (2020). *Mathematics for human flourishing.* Yale University Press.

Swanson, P. (2013). Overcoming the run response. *Mathematics Teaching in the Middle School, 19*(2), 94–99.

Tanco, M. (2019). *Count on me.* Tundra Books.

Tapp, F. (2023, October 19). *42 kid-friendly jokes about math and numbers.* Parents. https://www.parents.com/fun/math-jokes-for-kids-guaranteed-to-get-a-laugh/

Tapper, J. (2012). *Solving for why: Understanding, assessing, and teaching students who struggle with math* (1st ed.). Math Solutions.

TeachingWorks. (2023). *Leading a group discussion.* TeachingWorks Resource Library. https://library.teachingworks.org/curriculum-resources/teaching-practices/leading-a-discussion/

TEDxManhattanBeach. (2012, November 8). *The power of belief – Mindset and success* [Video]. Youtube. https://www.youtube.com/watch?v=pN34FNbOKXc&feature=youtu.be

The Nation's Report Card. (n.d.). *The nation's report card: 2022 NAEP mathematics assessment.* https://www.nationsreportcard.gov/highlights/mathematics/2022/

Vakharia, V. (2025). *Math therapy: 5 steps to help your students heal their math trauma and build a better relationship with math.* Corwin.

Van de Walle, J. A., Karp, K. S., & Bay-Williams, J. M. (2019). *Elementary and middle school mathematics: Teaching developmentally.* Pearson.

Wallmark, L. (2015). *Ada Byron Lovelace and the thinking machine.* Creston Books.

Wallmark, L. (2017). *Grace Hopper: Queen of computer code.* Sterling Children's Books.

Warshauer, H. K. (2014). Productive struggle in middle school mathematics classrooms. *Journal of Mathematics Teacher Education, 18*(17). Advance online publication. https://doi.org/10.1007/s10857-014-9286-3

Watson, C. M. (2022). Developing student agency as an act of love. *Mathematics Teacher: Learning & Teaching PK-12, 115*(7), 517–518.

Wedekind, K. O., & Thompson, C. H. (2020). *Hands down speak out: Listening and talking across literacy and math K-5.* Stenhouse.

Will, M. (2023, April 24). How much do teachers get paid? See new state-by-state data. *Education Week.* https://www.edweek.org/teaching-learning/how-much-do-teachers-get-paid-see-new-state-by-state-data/2023/04

Willis, J., & Willis, M. (2020). *Research-based strategies to ignite student learning: Insights from neuroscience and the classroom.* ASCD.

Winter, J. (2017). *The world is not a rectangle: A portrait of architect Zaha Hadid.* Beach Lane Books.

Woo, E. (2019). *It's a numberful world: How math is hiding everywhere.* The Experiment.

youcubed. (n.d.). *Cathy Humphreys teaching a number talk* [Video]. https://www.youcubed.org/resources/cathy-humphreys-teaching-number-talk/

youcubed. (n.d.). *Jo teaching a visual dot card number talk* [Video]. https://www.youcubed.org/resources/jo-teaching-visual-dot-card-number-talk/

youcubed. (n.d.). *Shape talks (3–5).* https://www.youcubed.org/wim/shape-talks-3-5/

youcubed. (n.d.). *Shape talks (k–2).* https://www.youcubed.org/wim/shape-talks-k-2/

Zager, T. J. (2017). *Becoming the math teacher you wish you'd had: Ideas and strategies from vibrant classrooms.* Stenhouse.

Zavala, M. D. R., & Aguirre, J. M. (2024). *Cultivating mathematical hearts: Culturally responsive mathematics teaching in elementary classrooms.* Corwin.

Index

**JENNIFER M. BAY-WILLIAMS,
JOHN J. SANGIOVANNI, ROSALBA SERRANO,
SHERRI MARTINIE, JENNIFER SUH,
C. DAVID WALTERS, SUSIE KATT**

Because fluency is so much more than
basic facts and algorithms.
Grades K–8

**MARIA DEL ROSARIO ZAVALA,
JULIA MARIA AGUIRRE**

Discover innovative equity-
based culturally responsive
mathematics instruction that
unlocks the mathematical
heart of each student.
Grades K–8

**IVANNIA SOTO,
THEODORE RUIZ SAGUN,
MICHAEL BEIERSDORF**

Focus on the literacy
opportunities that multilingual
students can achieve when
language scaffolds are
taught alongside rigorous
math and science content.
Grades K–8

**JOHN J. SANGIOVANNI, SUSIE KATT,
LATRENDA D. KNIGHTEN, GEORGINA RIVERA,
FREDERICK L. DILLON, AYANNA D. PERRY,
ANDREA CHENG, JENNIFER OUTZS, KAREN MESMER,
ENYA GRANDOS, KEVIN GANT, LAURA SHAFER**

Actionable answers to your most pressing
questions about teaching elementary math,
secondary math, and secondary science.

Elementary, Secondary

**CHRISTA JACKSON, KRISTIN L. COOK,
SARAH B. BUSH,
MARGARET MOHR-SCHROEDER,
CATHRINE MAIORCA, THOMAS ROBERTS**

Help educators create integrated STEM
learning experiences that are inclusive for all
students and allow them to experience STEM
as scientists, innovators, mathematicians,
creators, engineers, and technology experts!

Grades PreK–5 and Grades 6–12

A Sage Company

CORWIN HAS ONE MISSION: to enhance education through intentional professional learning.

We build long-term relationships with our authors, educators, clients, and associations who partner with us to develop and continuously improve the best evidence-based practices that establish and support lifelong learning.